# WHALING IN *Maine*

# WHALING IN *Maine*

Charles H. Lagerbom

Published by The History Press
Charleston, SC
www.historypress.com



First published 2020

Manufactured in the United States

ISBN 9781467144520

Library of Congress Control Number: 2020932162

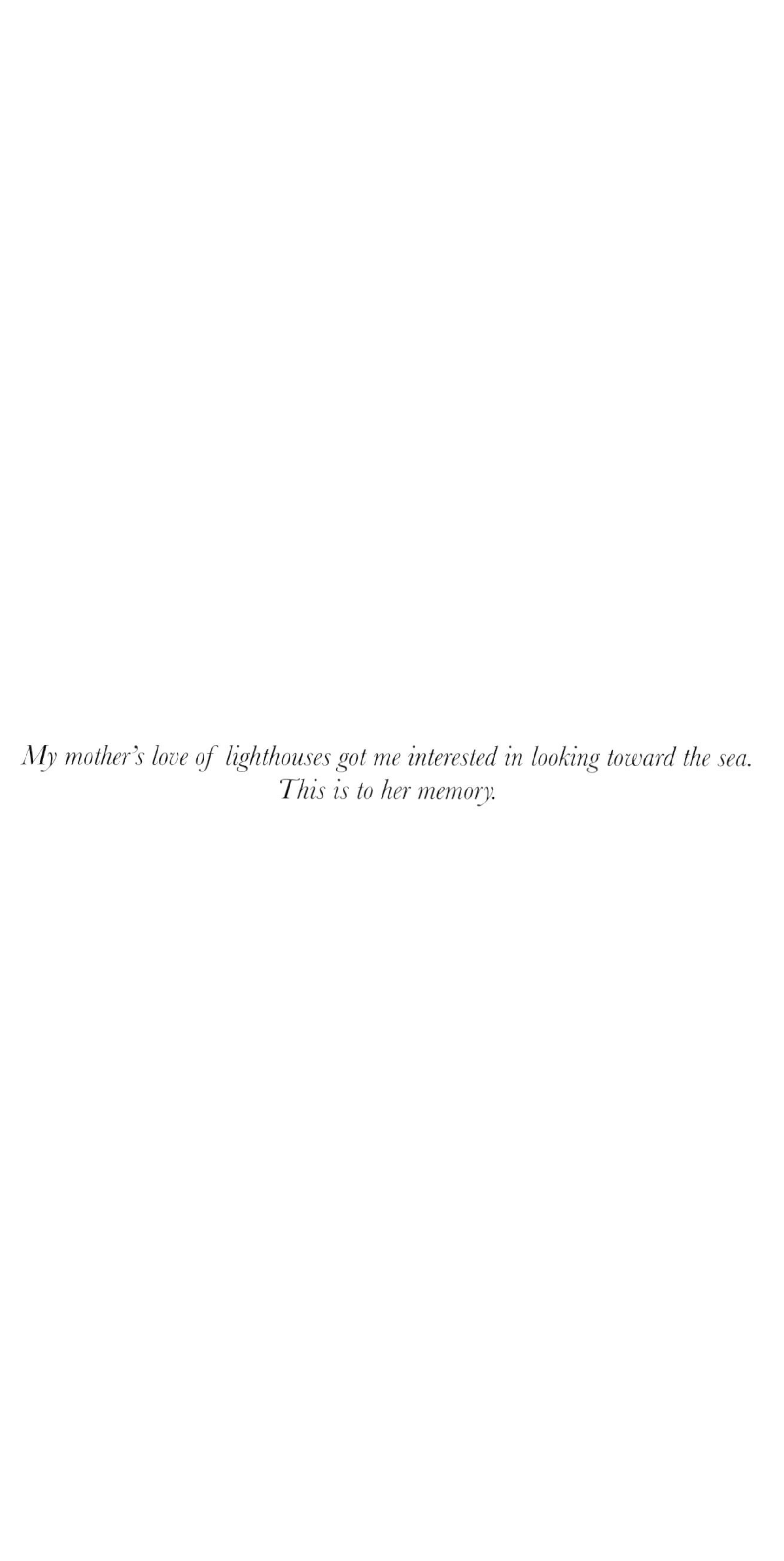

*My mother's love of lighthouses got me interested in looking toward the sea.*
*This is to her memory.*

# CONTENTS

# PREFACE

*The time has come to harpoon a whale.*
*—John Hope*

Whaling and the state of Maine have had a long and rich association. Even before Europeans arrived off its shores, native inhabitants were capturing whales that ventured near its rocky coasts from Gulf of Maine waters. There are stories of mythological beings of the Passamaquoddy and Penobscot Indians named Glooskap and Keanke interacting with whales. In one story, Glooskap gives a whale some tobacco to keep her happy. In another, they are featured spearing whales from a canoe off the Maine coast.[1]

As colonial migration and development began, Maine's association with its marine environment included the whale, as settlers developed limited, shore-based whaling, before venturing farther off-shore. No stranger to other fishery operations, Mainers also embraced the capture and harvest of cetaceans. When whaling became more of an industry, Maine and Mainers contributed to its development. The state and its people provided significant participation in whaling in many different ways, such as Maine-produced whaling equipment, Maine-inspired whaling gear inventions, Maine community whaling and sealing ventures, Maine-built ships that engaged in whaling, Maine-native captains of whale ships and Maine-grown crews who whaled in all corners of the globe. For the entire nineteenth century, until its eventual decline and demise, American

*Glooskap and Keanke Spearing the Whale. Collections of Maine Historical Society, courtesy of MaineMemory.net, item #7531.*

whaling operations often had some kind of Maine connection. Maine and Mainers were right in the thick of the industry with whale ships, whaling captains and whaling crews. It was a rich and unique historical era, one that helped formulate and entrench the iconic image of New England whaling. Maine was right there, all along the way.

It has been suggested that whaling was just a minor Maine fishery, especially compared to other more famous commodities like lobster or cod—that Maine's whaling history was nothing more than a passing interest or fancy, which often bore little fruit or success. This study seeks to challenge that assumption and suggests Maine's connection with American whaling was far more intertwined and integral to the overall industry than most realize or recognize, that it is an error to dismiss the wealth, diversity and richness of Maine historical ties with American whaling. Maine's involvement proved a vital source of energy, ideas, men, money and material, which helped transform whaling into the iconic, famous New England industry it became.[2]

Maine's initial involvement might appear modest. In the seventeenth and eighteenth centuries, besides processing drift whales that had died of natural causes and washed ashore, whales were mainly hunted just off the coast. Fairly easy to take, these whales were usually beached or easily corralled into shallow waters. Suitable boats were built to help guide these large beasts shoreward. But actually, catching and dispatching whales was done onshore—a purely local effort and result. A few Maine communities continued this kind of whaling into the nineteenth century, most notably Vinalhaven, Winter Harbor, Prospect Harbor and Mount Desert Island. These were limited ventures, however, mostly reliant on small boats and dependent on passing whales.[3]

It was not long before it was required to send larger, better equipped boats out farther to catch them in open sea. Between 1789 and 1815, the average size of whaling ships increased to three hundred tons. No longer boats, these vessels became true whaling ships in that they could now process whales on capture, rendering them into oil for barrels, all the while in open sea far from their home port. New Bedford, Fairhaven and New London in Connecticut took lead in this, but it proved a challenge for Nantucket, which had been at the forefront of early whaling. These larger, deeper vessels were challenged by Nantucket's shifting sand harbors, which allowed New Bedford to eventually replace it as a premier whaling center.[4]

Maine got involved with these larger whaling ships and longer whaling voyages. Maine community leaders expressed interest in these more ambitious, yet still somewhat limited, expeditions. They proposed, financed and manned some community-wide enterprises as business ventures. Tiny, isolated coastal towns built their own vessels and ventured far afield into Greenland and Arctic waters after seals, fish and whales. These proved financially rewarding, although extremely dangerous, as loss of a vessel and crew could financially, demographically and emotionally devastate an entire community.

These ventures also proved good training for Mainers as crew and eventually captains and influenced ship construction in Maine shipyards. "Every ship's deck was then a pathway to honorable distinction, but in no other ship could an ambitious green hand gain rank so quickly as on a whaler; for the whaler carried more officers than any other ship, and the necessities of the work compelled the officers to give much attention to the training of the inexperienced."[5] Maine provided numerous seasoned crewmen and captains to the growing industry, many of whom gravitated to Nantucket, New Bedford, New London, Fall River and other principal whaling ports. This was aided by development and usage of the lay system, where a young man worked for a percentage or share of the catch. A Mainer could do well if he shipped aboard a profitable whaling ship.

Maine did not just contribute men. Near the end of the American Revolution, American whalers were indignant with British whalers making contractual deals with inhabitants along Maine's Kennebec River. Captain Alexander Coffin wrote to Samuel Adams about his British competitors contracting with Kennebec settlers to provide annual supplies of hoops, staves and other lumber necessary for whaling. Adams prompted Massachusetts officials to get involved, and the deal was forbidden. By the early 1800s, those same wood resources from Kennebec valley, such as barrel staves, spars, oars and stores, were annually going to Nantucket and New Bedford for their whaling fleets. Nantucket whaling captain John O. Morse carried huge hand-hewn beams of red pine from Maine for Nantucket's Edgartown Methodist Church on his whaling bark *Rhine* on its December 1842 voyage. That particular voyage returned September 1845 with the red pine beams and four hundred barrels of sperm whale oil.[6]

*Sperm Whaling with Its Varieties. Library of Congress.*

Belfast's newspaper, *Republican Journal*, reported in 1831 that a Waldo County man named David Pierce had secured a contract to furnish all oars used by whalemen of New Bedford and New London. Pierce was dubbed the "Oar King" and received sixty dollars per thousand feet. He reportedly shipped over one hundred thousand feet per annum from Montville, where he fashioned the oars by hand from local ash. A set totaled seven oars, including a steering oar, and two each of twenty-two-foot, twenty-foot and eighteen-foot oars. They were reported as very handsome and serviceable.[7]

Even Aroostook County, known more for farming and potatoes, had whaling connections. Southern Aroostook Agricultural Museum in Littleton holds a whaling blubber knife from 1900. A sperm oil lamp dated to 1790 can be found at the Aroostook County Historical Art and Museum. Used to burn sperm whale oil, these were preferred household lamps as they burned cleanly without producing offensive odors.[8]

At least one Mainer dabbled in scrimshaw art. His Maine origins referred mistakenly to Pittsford, but Sylvanus Baker Jr. was a native of Pittsfield, Maine. Referred to as the Pagoda Artisan, Baker was known for his pagoda structures that adorned his whale tooth carvings. He was thirty-five years old as a boat-steerer and third mate aboard bark *John Dawson* on its October 1855 voyage to the Atlantic and Indian Oceans under Amos Crowell Baker Sr. Though it is not clear if they were related, Sylvanus Baker signed aboard for 1/55th lay. They returned in May 1859 with 577 barrels of sperm whale oil. Three of Sylvanus's carvings are on display at New Bedford's Whaling Museum.[9]

In 1880, many Maine fishermen temporarily turned to whaling in the Gulf of Maine when their menhaden industry collapsed. Known as porgies, pogies, bony-fish, mossbunkers, bug fish, fat-backs, whitefish, bunkers, old-wives, chebogs or greentails, these middling-size fish were dominant prey for many larger predatory fish. Filter feeders much like baleen whales, they produced oil, which was used for tanning and curing leather. Prolific and easy to catch, menhaden was a huge industry. It is said that even when the whaling industry reached its height, it was still outweighed by menhaden oil. But for about six years starting in 1880, this middle fish in the food chain completely disappeared from the East Coast north of Cape Cod.[10]

Using newly developed steam engines, menhaden fishermen turned to finbacks, humpbacks and blue whales spotted in the Gulf of Maine. But their steamers were ill-equipped and smaller than typical whale catcher boats. Boothbay's Luther Maddocks began hunting humpbacks off

Carved teeth from the Pagoda Artisan, Pittsfield native Sylvanus Baker at the New Bedford Whaling Museum. *Photo by author.*

Monhegan Island and, in 1885, made money by exhibiting a whale carcass in Portland in conjunction with the Nineteenth National Encampment of the Grand Army of the Republic. Other menhaden fishermen who turned to whaling included Albert Murray, who hunted from steamer *Fanny Sprague*, and brothers Alonzo, Stephen and Arthur Nickerson. There was also Major W.S. White of Rockland, Joshua G. Nickerson, William G. Butman and General Davis Tilson. Vessels included *Mabel Bird*, built in Bath and registered at Portland, and Rockland's *Hurricane*, built and registered there. Operations centered in East Boothbay; Linekin's Bay; Carver's Harbor on Vinalhaven; and islands such as Monhegan, Greene and Long. But the move was temporary, never profitable and often inefficient. Menhaden fishermen often lost their catch, and dead whale carcasses caused local sensations when they washed ashore. By 1889, the menhaden had returned in sufficient numbers, and the short diversion to whaling was pretty much over.[11]

By far, Maine's most significant connection to American whaling was its shipbuilding, as evidenced by numbers of vessels from Maine shipyards, which went directly or indirectly into whaling. When tonnage size of whaling vessels increased after 1815, Maine was well positioned to build and launch these larger ships. Whale ships often changed to bark rigs with square-rigged yards and sails on their mizzenmast replaced by fore and aft sails. This configuration increased maneuverability, especially among shallows and ice floes of northern or southern whaling and sealing grounds. Such a move also decreased manpower needed to sail these vessels, which allowed crew numbers, and therefore shares of catch, to remain optimal.[12] As American whaling became an industry, Maine was right there to do its part.

# ACKNOWLEDGEMENTS

In addition to my wife, Jennifer, and children, Charlie and Audrey, I am grateful to the following, who helped me with this research project: John F. Berube, Andover High School (Andover, Massachusetts); Michael Bloomfield, National Archives and Records Administration (College Park, Maryland); Richard Brockman, Maine Maritime Museum (Bath, Maine); Cameron Camara, LaFrance Hospitality (New Bedford, Massachusetts); Anne Farrow, Maine Maritime Museum (Bath, Maine); John Golden, Penobscot Marine Museum (Searsport, Maine); Cipperly Good, Penobscot Marine Museum (Searsport, Maine); Kevin Johnson, Penobscot Marine Museum (Searsport, Maine); Michael G. Kinsella, The History Press (Charleston, South Carolina); Nathan Lipfert, Maine Maritime Museum (Bath, Maine); Chris Maxworthy, Australian Association for Maritime History (Australia); Lesley Martin, Chicago History Museum (Chicago, Illinois); Kelly Page, Maine Maritime Museum (Bath, Maine); Kay Peterson, Smithsonian Institution (Washington, D.C.); Megan Pinette, Belfast Historical Society and Museum (Belfast, Maine); Mark D. Procknik, New Bedford Whaling Museum (New Bedford, Massachusetts); Sarah Timm, Maine Maritime Museum (Bath, Maine); A. Bowdoin Van Riper, Martha's Vineyard Museum (Vineyard Haven, Massachusetts); and Sofia Yalouris, Maine Historical Society (Portland, Maine).

The following organizations were also very helpful regarding access, research, images, artifacts—pretty much all aspects of this project. Their assistance took many forms, all greatly appreciated. Thank you to: Penobscot Marine Museum, Maine Maritime Museum, Maine Historical Society, Belfast Historical Society and Museum, New Bedford Whaling Museum and National Archives and Records Administration.

Introduction

# A WORLD OF WHALING

*Where great whales come sailing by, Sail and sail, with unshut eye,*
*Round the world forever and aye.*
*—Matthew Arnold, "The Forsaken Merman"*

American whaling industry peaked in 1846 with 736 registered American whaling ships and for the next five years, averaged 638 American vessels. By 1850, most American whalers outfitted and crewed for multiyear voyages. General destinations included any and all of the known whaling grounds, including "off-shore" grounds first discovered by George W. Gardner in 1818. Early on, Atlantic waters had been productive with one whaling area stretching from South America to subantarctic islands over to Africa's Cape of Good Hope. Far north, there was Hudson Bay, Davis Strait and Arctic waters, restricted mainly to late spring and summer. British whale ships headed home after each season, but American whalers began to venture south instead, whaling until northern waters reopened the following season. This routine kept American vessels at sea for multiyear voyages, usually between two and five years. Being out for so long, whalers often charted other ships to transport home early loads of oil and baleen. These were either unsuccessful whalers with room or convenient merchant ships bound stateside. This allowed whalers to remain longer on station and brought even more vessels into the industry.[13]

By 1800, New England whaling operations were rounding Cape Horn and Cape of Good Hope. Valuable grounds were found in southern Indian

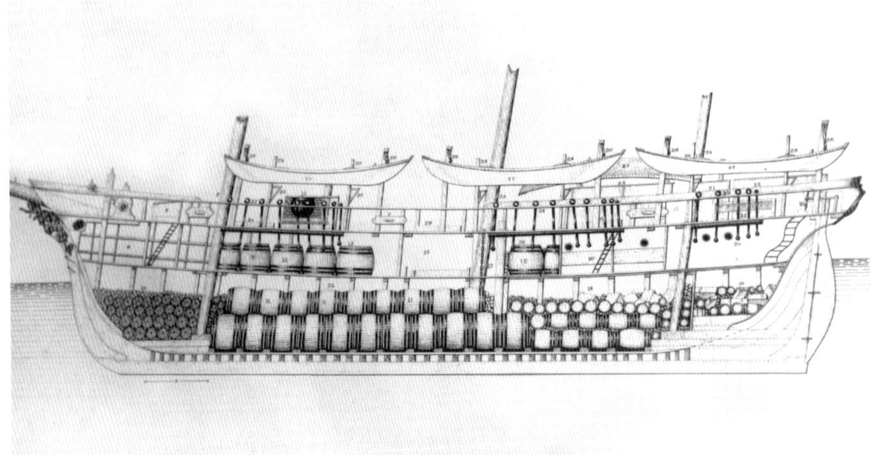

Interior profile of a whaling ship. *From the Fishing Industries of the United States, 1889.*

Ocean waters around Heard and Kerguelen Islands, which were rich in whales, seals and sea elephants. South Africa's Cape Town became a necessary stop. Pacific whaling grounds were found off Australia and New Zealand, as well as farther south on islands off Antarctica. Other grounds included Japanese waters, the Okhotsk Sea, the United States' northwestern coast and the Bering Sea, which opened into very far northern waters off Point Barrow, Alaska. Many fleet vessels spent their seasons in Arctic waters and then made their home port in Honolulu or Lahaina in the Hawaiian Islands. In 1822, over 60 whaling vessels stopped at Honolulu to replenish food and water supplies. In a few years, over 140 ships were stopping there annually. In 1846, almost 600 ships arrived at Lahaina or Honolulu.[14]

From its beginnings, American whaling's primary objective was to fill barrels with as much oil as quickly as possible, ideally from whales, seals, walruses or sea elephants. Whalers went after three different raw materials: oil, spermaceti oil and whalebone. Oil was a primary lubricant for machinery. Spermaceti oil, solely from the head-case of sperm whales (*Physeter macrocephalus*), was more expensive than other whale oil and was used for illumination by burning cloth wicks in lamps or by candles. It was prized for its clean-burning properties. Sperm whale oil was more sought after, although regular oil from other whales and seals also paid bills. Whalers also took whalebone, mostly baleen, a filter-feeder system inside mouths of baleen whales made of keratin, similar to human fingernails and hair. Baleen bristles are arranged in plates across the animals' upper

*Right*: Sample of spermaceti oil at New Bedford Whaling Museum. *Photo by author.*

*Below*: Baleen hairpin, circa 1840. *Collections of Maine Historical Society, courtesy of MaineMemory.net, item #48948.*

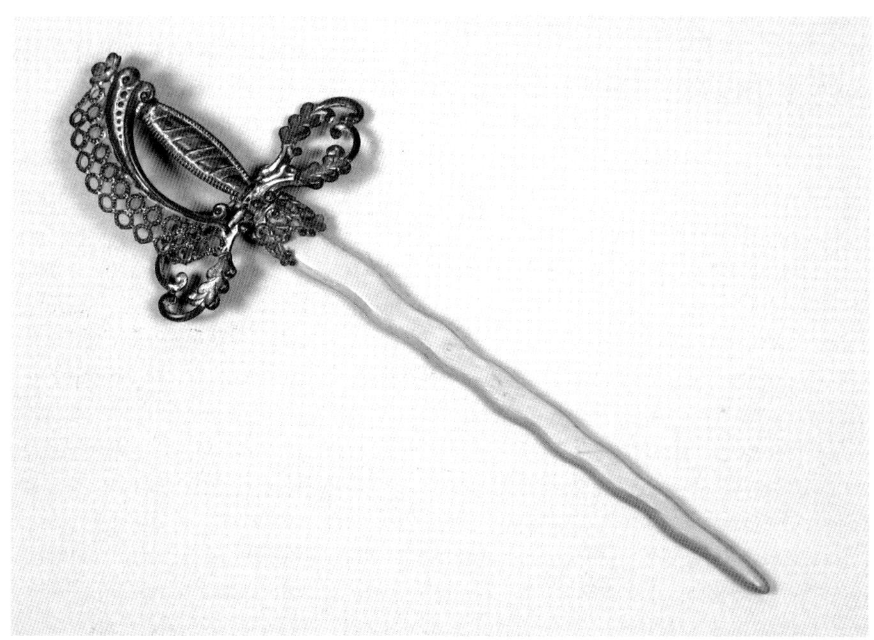

jaws. Whalers brought back tons of baleen to be fashioned into buggy whips, ribs of parasol umbrellas, petticoats and corsets. Ivory and bone from whales, such as sperm whale teeth, were often engraved with images or fashioned into tools and other assorted items. In their collections, the Maine Historical Society has a woman's ornamented hairpin in the form of a sword made from baleen. There are many such items scattered around the state. Public records of United States exports of oil, sperm whale oil and baleen date as far back as 1791. These iconic New England whaling products and the items fashioned from them represented a major portion of American gross domestic product for nearly a century.[15]

One reason whaling companies did not develop in Maine prior to 1834 was the high cost of outfitting these special vessels combined with the high risks and potentially low returns. A decade after its statehood in 1820, Maine whaling ventures started to form in coastal communities. Before, the state had lacked the cooperative collective spirit necessary for group specialization. Maine ports had also been engaged in developing other types of maritime commerce. Their geographic remoteness from Nantucket and other early whaling centers was also a factor. But between 1834 and 1843, that all started to change.[16]

I.

# COMMUNITY VENTURES, RISKS AND REWARDS

*They that go down to the sea in ships; and occupy their business in great waters; these men see the works of the Lord; and His wonders in the deep.*
*—Psalms 107:23–24*

Lured by profits, early nineteenth-century Maine coastal communities built vessels to send their men to Greenland whaling and sealing grounds. It was expensive and dangerous but held the promise of large financial rewards. They outfitted and crewed these ships with locals in true community-wide efforts. If successful, the windfall could be enormous. However, an unsuccessful or devastating trip, had the potential to decimate their local economies and village demographics for decades. It was high risk and high reward of the highest order. In the early 1800s, the Cranberry Islands near Mount Desert Island were more heavily settled than Bar Harbor. Voyages to northern whaling and sealing grounds had already been successful. It was exciting for young men to see strange cargoes brought back, such as carved bone, skins, native spears and harpoons and clothing of fur and feathers, as well as stuffed seals, bears and birds. Samuel Hadlock Jr. took a locally built and crewed ship named *Minerva* to Greenland but never returned. *Minerva* was a small, two-masted schooner built in Cranberry Pool in 1820. Sixty feet long and only seventy-five tons, it was owned by Hadlock's father and made an eight-month voyage to Greenland in March 1829. With one deck, it barely held a crew of seventeen in addition to Hadlock Jr.

as captain. His crew members were Mount Desert and Cranberry Island locals, representing significant percentages of young men from the area. Mostly in their teens and twenties, they included William Gilley, Joseph Ober, Joseph Stanley and Ellis Kingsbury, as well as Hadlock's fourteen-year old neighbor. Brimming with promise of wealth and adventure, as well as trust in Hadlock's experience, they eagerly joined the proposed eight-month voyage. Ship and crew were never seen again. Similar in tragic outcome but with even less historical record about it, a man named George Dix and a schooner named *Esther Linda* were lost on a similar venture to Labrador.[17]

During the 1830s, some Maine communities watched with growing interest as the whaling successes at Nantucket and New Bedford flourished. Wanting in on profits, they thought, why not Maine? More communities did not jump in because Maine shipbuilding enterprises were maximizing on early mast trade and then flourishing East India trade. Already strong in fisheries, lumbering and ice industries, it is argued that Maine just did not have enough manpower or investment capital for whaling. Still, some Maine communities were ready to try, including Portland, Wiscasset, Bath and Bucksport. Some proved more successful, if only for a brief period of time, as most were finished within a decade or so. But individual efforts led the way.[18]

In 1833, Mainer Charles Cushing of South Berwick went into whaling. He purchased and outfitted the ship *Triton* and sailed from Portsmouth, New Hampshire, to the south Atlantic under master Richard Flanders. They quickly returned with 2,000 barrels of oil, 450 of which were sperm whale. The value cashed out at over $500,000 by today's measure. *Triton*, quickly refitted, departed in May 1834 under master Daniel Flanders and returned in April 1835 with 1,400 barrels of oil. Later, William Ritchie sailed *Triton* in July 1835 but returned in 1837 empty. By then, Cushing had bought *Plato*, which sailed under master Benjamin Manter and returned in February 1835 with 950 barrels of oil, 250 of which were sperm whale, as well as seven thousand pounds of baleen. Cushing closed out his business, took his profits and quit just before the economic Panic of 1837. Charles Cushing drowned April 22, 1872, and was buried in Evergreen Cemetery in Portland, Maine. *Plato* went to New Bedford but was grounded. In 1842, *Plato* wrecked off Montauk, New York. *Triton* was condemned and broken up in 1849.[19]

Portland and Wiscasset are considered the first Maine municipalities to engage in commercial whaling. In 1827, 388-ton *Science* was built in Newbury,

The Greenland Whale Fishery, panel one, by Joghem de Vries, circa 1769. *Penobscot Marine Museum.*

Massachusetts. In the 1830s, Portland acquired it for its venture. Town financiers arranged purchase and recruited trained crew from southern New England to sail it. On its main deck, *Science* was outfitted with large brick tryworks for boiling down oil. It sported four thirty-foot whaleboats, three of which could be put into water the moment whales were spotted. *Science* departed Portland for Pacific in January 1834, with master Captain Alexander Whippey and managing owners Chadwick and Davis. Whippey took it west of Chile and Peru, then plied south Pacific waters, sailed to Japan and finished off in New Zealand. After two years, it took 500 barrels of oil and then fell in with New Bedford whaler *Canton* under master Abram Gardner. It was a long, grueling voyage with the loss of one man, Amos Burnham, overboard and several desertions when they stopped at Hawaii to resupply. *Science* returned on May 4, 1838, with 2,100 barrels of sperm whale oil.[20]

Its long voyage cut deeply into profits, so Portland refitted it for another trip, and it sailed September 8, 1838, with Whippey as master. Caleb Adams was listed as managing owner or agent. *Science* was seen that December already having taken three whales. It returned June 4, 1841, with 3,100 barrels of oil, 300 of which were sperm whale. But *Science* returned in difficult economic times. The Panic of 1837 convinced Portland owners that they could no longer compete with New Bedford or Nantucket whaling. They decided to quit the business, and *Science* was sold to New Bedford owners, who made one more voyage under master William H. Wood to the northwest coast of America, which netted 2,944 barrels of oil, 180 of which were sperm whale, and almost seven tons of baleen. *Science* was sold in 1847.[21]

In 1834, Wiscasset conducted its own whaling venture with the 380-ton ship *Wiscasset*, built in Maine in 1833. Hoping to cash in on lucrative stock corporations in whaling, like New York river towns Hudson or Poughkeepsie, Wiscasset residents pooled resources and formed Wiscasset Whale Fishing Company with William M. Boyd as president and William Wood as secretary. Twenty local shareholders and a state legislature provided its charter, and agents purchased *Wiscasset* for $20,000. Interest was fanned by local newspapers, which called attention to potential profits of such an enterprise. To outfit it, whaling gear was purchased and shipped from New Bedford, and Nantucket captain Richard Macy was master, and Jonathan Parson was managing owner. Enough locals were recruited to sail it to the Azores, where there was more seasoned crew. *Wiscasset* outfitted at Parson's Wharf on November 29, 1833—an occasion celebrated by renaming the site Whale Ship Wharf. *Wiscasset* sailed down Sheepscot River on May 13, 1834, and successfully rounded Cape Horn, although its rudder was damaged on the stormy passage. When it was spotted in April 1836 off New Zealand, *Wiscasset* had already taken 1,350 barrels of oil.[22]

Off New Zealand in 1836, *Wiscasset* captured a sperm whale that produced ninety barrels of oil. Sailors kept her jawbone and whale teeth. One tooth, engraved with *Wiscasset*'s likeness, was inscribed "A tooth of a 90 bbl whale got on the coast of New Zealand, Jan. 17th, 1836." Kept by Marine Society in New York, it was presented to Andrew Carnegie on July 13, 1903, and as of this writing is on display at Kendall Whaling Museum in Sharon, Massachusetts.[23] *Wiscasset* returned from that three-year voyage on September 10, 1837, with 2,880 barrels of sperm whale oil. It was successful, although the crew lost one man overboard. Word of its success reached Wiscasset ahead of the ship's return, and townspeople turned out for a celebration. The souvenir jawbone was tied up wharf-side and remained there for years. All twenty stockholders received dividends, ship and crew all paid from that one voyage.[24]

*Wiscasset* went to the Pacific again on January 27, 1838, with master Seth B. Horton and John Brooks as managing owner. Horton faced crew desertions at Tahiti, as well as arson, when the cabin boy tried to set the ship afire, forcing the crew to return to Tahiti for repairs. It returned on July 22, 1841, with 2,100 barrels of oil, 900 of which were sperm whale, and had sold 600 barrels in Bahia. The crew even brought back 150 pounds of coffee

Postcard of *Wiscassset* ship. *Maine Maritime Museum.*

for side trading. But with economic depression underway and rising prices of whaling gear and supplies, *Wiscasset*'s owners decided to sell out. One final reminder of Wiscasset's whaling days is a weathervane perched atop the town's municipal building. Made by Gil Whitman, it is patterned on Wiscasset's whaling ship. *Wiscasset* was sold to Sag Harbor owners and made two more voyages to New Zealand and the northwest coast of America before being withdrawn and sold in 1847. It served as a transatlantic packet for New Line of New York. On one of its crossings, it brought a young immigrant named Andrew Carnegie to America. When he was presented with a scrimshaw piece from *Wiscasset*, he remembered how it "rolled badly, was slow, and uncomfortable." Carnegie's private yacht was 380 tons larger.[25]

In 1841, Bath decided to try its luck, mostly at the urging of a China, Maine native named David Densmore. At thirteen, he had traveled to Pittston and saw his first sailing vessel. Talking with a China neighbor who had once been whaling inspired him to go to Nantucket. Densmore was nineteen years old.[26] He sailed aboard *Peru* in 1833 and later wrote a readable account of his life. His autobiography, published forty years later, recounted his first encounter up close with a whale:

> *I looked in every direction, my eyes as big as demijohns. Directly I heard a roaring sound. The whale had come up, and the sound I heard was the noise he made in spouting…the boatsteer let fly first one harpoon and then the other.…The whale was a large one, and seemed decidedly opposed to being interfered with in that unceremonious matter. He kicked and thrashed round furiously…Finding he couldn't get clear by kicking with his enormous flukes, or tail, he commenced running to windward at a furious rate, dragging the two boats after him.*[27]

By 1833, Densmore was *Peru*'s third mate and even brought one of his brothers along. He returned to China long enough to marry and enjoy a six-day honeymoon and then shipped as second mate aboard *Henry*. After this, he tried his hand at farming. By 1841, Densmore was ready for sea again and moved to Bath. It was then, he wrote later, that he heard an inner voice inspiring him to form his own whaling company. Densmore was able to raise funds by selling subscriptions to wealthy Bath residents. Under his guidance, subscribers charted Bath Whaling Company with thirty-eight members. A vessel was then purchased—206-ton *Massasoit*, built in Plymouth, Massachusetts, in 1824 or 1825. It was altered to a brig. When furnishing it, Bath Whaling was fortunate that Portland's earlier venture with *Science* was ending, and the company bought *Science*'s gear for half price at auction.[28]

*Massasoit*'s captain, Densmore, now twenty-eight years old, oversaw its fitting out, which dragged through October and November 1841. When *Massasoit* did sail, it was with a motley crew of thirty-four either drunken or greenhand sailors. Most were foreign, although some came from Densmore's hometown, including his brother, who was first mate, and his cousin as third mate. Another brother and another cousin signed on as seamen. Three weeks out, Densmore faced a potential mutiny from some crew who were unwilling to round Cape Horn. Densmore grabbed a pistol, called all hands on deck and gave what he termed a "stump speech," with dire consequences for any insubordination. Either the speech worked, or he had misunderstood the rumblings, but *Massasoit*'s crew offered no resistance to orders. Two men grew sick and died. By May 1842, they had captured nothing. In July, they put in at Fayal with 120 barrels of sperm whale oil, and by the end of the first year had taken 230 barrels of oil. Homeward bound in February 1843, it only had 300 barrels, but the crew managed to find 100 more off Cape Hatteras. Another inner voice came to Densmore off Cape Cod—a passing vessel had warned of a coming nor'easter and suggested *Massasoit* sail offshore. Densmore's inner voice told him the opposite, so he tacked inshore. The storm became southwesterly and wrecked ships that had stayed out to sea, including the one that had warned Densmore. The inner voice also encouraged him to conduct religious services onboard the whale ship.[29]

David C. Densmore frontispiece. *From* The Halo: An Autobiography, *1876.*

In the Kennebec River in March 1843, *Massasoit* grounded on rocks just below Fiddlers Reach and had to be lightened of oil and tryworks before being floated off. It was an anticlimactic end to a meager voyage. Densmore arrived in Bath, where he found interest in whaling ventures diminished, although he claimed he had not lost money and may have been rationalized by including insurance payments and money from reselling *Massasoit*. When Bath Whaling Company folded, Densmore went on to other careers, such as

Religious service aboard a whaler. *From* The Sea and the Sailor, *1851.*

mill operator, Rockland shipbuilder, California forty-niner and cobbler. By the 1860s, he was a professional occultist, spiritual healer and writer. He also owned the *Voice of Angels* newspaper, which published messages from beyond. Densmore's autobiography was published in 1876.[30]

Also in 1841, Bucksport outfitted schooner *Warwick* for its own whaling venture.[31] It sailed in April under a master named Grozier but did not get far before facing a storm necessitating the crew's return. While laying to, *Warwick* had its "decks completely swept of everything except the tryworks, lost three boats, caboose, and house binnacle, booby hatch, oars, whaling craft and mainsail. She had a crew of 15 men, nearly all of whom were washed overboard but saved as if by a miracle, one man having caught a piece of rigging, by which he regained the vessel, and the others being washed back upon her decks by the returning wave."[32]

After repairs, *Warwick* put to sea and sailed to the Azores, catching blackfish, small whales that give decent oil, which the crew sold in the islands. It returned to Bucksport in April 1842 with only 110 barrels of sperm whale oil. There was no further Bucksport effort, and *Warwick* was sold to a Boston firm as a merchantman.[33]

Bangor outfitted *Beulah* in 1838 for a whaling venture, but records are scarce, and its master, destination and catch results are unknown. Portland may have tried again, but the only reference to that is a journal kept aboard *Elbe* from 1845 to 1848 while out of Hamburg, Germany. Kept by sailor George F. Neil, it refers to *Elbe* as a Portland vessel, although whether he meant Portland, Maine, or England is not clear. The only *Elbe* recorded was from Poughkeepsie, New York, and was lost in Cook's Straits in December 1841.[34]

Individual Mainers purchased ships and either kept them in whaling or made other uses of them. In 1855, Portsmouth built 96-ton schooner *Washington Freeman*. It sailed from Fairhaven on three whaling voyages to the Atlantic, starting in 1867. Its two-year voyage, starting in 1868, under master Loring Braley included James Riley of Eastport, Maine. The crew brought back 470 barrels of oil, 158 of which were sperm whale. In 1871, it was sold to owners in Thomaston and put into freighting. In 1864, Fairhaven built 262-ton schooner *Glacier*. It became part of New Bedford's fleet and operated in Hudson Bay and Cumberland Inlet. Its 1866 voyage under master Edwin A. Potter included Mainers Frank Lailer of Bristol and Augustus Chadwick and John Keene of Augusta. *Glacier* brought back 220 barrels of oil, only 20 of which were sperm whale, as well as two tons of baleen. It was sold to Wiscasset owners in 1873. The 127-ton schooner *Georgia* made two whaling voyages to Atlantic waters from 1866 until 1869 under master Ebenezer Bradbury Jr. Even though two seasons yielded 235 barrels of whale oil, all but 4 sperm whale, the voyages were not deemed successful enough, and *Georgia* was sold to owners in Brewer, Maine.[35]

2.

# MAINERS

*...for a whaleship was my Yale College and my Harvard.*
*—Herman Melville, Moby-Dick*

Maine's proximity to growing whaling centers like Nantucket, Salem and New Bedford attracted young Mainers. Prior to the 1830s, whaling ventures rarely produced anything official in way of crew lists. While records are sketchy for the eighteenth century and earlier voyages, many likely had Maine connections. According to one source, many Mainers likely served aboard these whale ships, but their involvements were often considered "off the radar."[36]

One example is Mainer William Lowder. Born 1785 in Castine, he followed family tradition into maritime trades, later becoming shipwright and ship captain. Some records indicate that he went whaling, rounded Cape Horn and pioneered routes into Pacific waters. New England whalers were making multiyear voyages there since 1791, while other ships sailed for China, such as *Massachusetts* under Amassa Delano of Duxbury, Massachusetts. Its 1790 voyage included carpenter's mate Elias Parkman, who later resided on Penobscot River in Maine and may have been an influence on young Lowder. Another possible link was wealthy Bostonian and Lowder in-law Benjamin Bussey. He bought a house on Boston's Hawley Street that was once owned by Joseph Barrell. With others, including Bussey, Barrell outfitted Boston's first vessel to sail around Cape Horn. Their merchant connection may have produced the opportunity for William to go whaling.[37] But when and on what ship?

National Maritime Digital Library lists eighty-one American whaling voyages to the Pacific between 1797, when William was twelve, and 1804, when he was nineteen. Only eight also listed their destination as Chile or Valparaiso, a location Lowder said he had visited. Of those eight, one was out of Boston, six were from Nantucket and one was from New York. In 1798, *Active* sailed under Nantucket-born Micajah Gardner, perhaps with Lowder aboard. In November, Gardner visited Desterro, present-day Florianopolis in southeast Brazil, lacking water and provisions. He remained twelve days and then rounded Cape Horn to Chile. *Active* returned to Boston in 1799 with 4,900 barrels of oil, 2,500 of which were sperm whale. Or Lowder may have shipped aboard Nantucket whalers to Chile, like *Hope* from 1797 to 1799 under David Giles, *Olive Branch* 1797 to 1799 under Obed Paddock, *Union* in 1798 under Grafton Gardner, *Minerva* 1798 to 1799 under Jared Gardner, *Lion* 1802 to 1805 under Peter Paddack or *Harlequin* 1803 to 1804 under Levi Starbuck. Lowder sailed *Goodrich* to Greenland in 1807, although that vessel is not listed in the maritime database. Perhaps in error, William Lowder, in his later years, was thought to have been whaling in Pacific waters, although his name does not show up.[38]

An official act of Congress on February 28, 1803, changed this hit-or-miss approach to crew lists and provided the first legal requirement to keep crew lists as part of ship's official papers:

> *Before a vessel could depart on a foreign voyage, the master had to deliver a list of the crew, verified by his oath, to the customs collector at that port. The collector then supplied the master with a certified copy of the list, copied in a uniform hand, along with a Clearance Certificate, at which time the master entered into a four-hundred-dollar bond to exhibit the Crew List to the first boarding officer he encountered upon his return to a U.S. port. There he was required to produce the persons named and described in the Crew List to give account for any crew members who were not present. Notes certifying sickness, discharge or desertion, usually signed by a consular official, were often included with the original list in order to prove that individuals not present were legally accounted for.*[39]

These lists, all recorded at customs houses in Fall River and Salem, as well as New London, Connecticut, represent hundreds of whaling voyages. Lists of crews also came from records kept by chaplains of New Bedford Port Society. The result is an extensive database from hundreds of vessels, hundreds of voyages and thousands of crew members in a searchable

William Lowder family grave site. *Photo by author.*

database that includes over 170,000 entries. This list represents only men who left home port and not those who joined later. Some mention desertions or deaths, but these are incomplete and are likely records of what chaplains read in newspapers and added to their files. Thirty-three states and two U.S. territories are represented, but the vast majority list no birthplace or residence. Of those that do, over 2,100 Mainers appear who served on nearly 620 different whaling ships between 1825 and 1910. These men represent .0124 percent of the total database and are only a fraction of the Mainers who likely went whaling.[40]

So, who were these young Mainers who went whaling? While it is impossible to identify them all, online databases from National Maritime Digital Library and Whaling History's website reveal that Maine contributed a larger share of whaling crew than one might think. Another way to find them is a search through local newspapers from the 1840s and 1850s, which are filled with notices of Maine men involved in whaling, often with bad news. One example was a notice of loss of all hands of brig *Thule* on its fourth whaling voyage, which had Mainers Alonzo O. Hatch of Kennebec and John Huntley aboard. *Thule* sailed from Nantucket in June 1842 under

Charles N. Coffin and was lost in the South Pacific in September 1844. Other examples contain notices of those killed by whales, such as eighteen-year-old D. MacDonald of Eastport, Abraham Stubbs (or Butts) of Portland, boat-steerer John D. Clark of Augusta and twenty-one-year-old Ethan Forbush of Sanford. Yet other notices mention those who died of sickness while whaling, such as Alvan Robinson of South Norridgewock, who died after suffering for two weeks with dysentery.[41]

It has been suggested that another way to measure Maine's whaling involvement is to note the disproportionately large amount of scrimshaw in the state. Maine Maritime Museum in Bath has a large collection of scrimshaw and ivory items and tools such as mallets, belaying pins, letter openers and bone handles for hacksaws and chisels. It also has a bone swift on exhibit. A swift is a complex tool that expands to hold open a skein of yarn and then rotates so the knitter can wind the yarn into a ball. Penobscot Marine Museum in Searsport, Maine, also has a large collection. One of its pieces is from whaleship *John Coggeshall*. It is a sperm whale tooth with the ship's likeness carved on it. Two Mainers shipped aboard its 1855 voyage, first under Phineas Fish II, who died at sea and was replaced by Caleb R. Lambert. Charles Harris of Portland and Asa Thompson of Chesterville returned in May 1859 with 1,261 barrels of oil, 112 of them sperm whale, as well as over six tons of baleen. Mainer George Maxfield of Shapley shipped aboard it in 1860 under Aaron Dean and returned in 1864 with 2,188 barrels of oil, 164 of them sperm whale, and sixteen tons of baleen. *John Coggeshall* was sold to New York owners and then to foreign owners that same year.[42]

Very few Maine men are identified before the 1840s. Earliest entries include August 9, 1825, when twenty-six-year-old Ellsworth-native William F. Trueworthy shipped from Fairhaven aboard *Amazon* under Ambrose Whiteus. The crew returned with 800 barrels of oil, only 100 sperm whale, and three tons of baleen. There was also twenty-one-year-old Isaac Tibbets of Sanford, who shipped aboard *Java II* on its first whaling voyage on November 5, 1828, under Barzallai S. Adams. The ship returned two years later with 2,040 barrels of oil, 120 of which were sperm whale, and eight tons of baleen. Two others, likely related, served on *Chelsea* on its first voyage out of New London on August 21, 1828, under James Davis. They were Portland's Robert Smith and forty-one-year-old George W. Smith from New Gloucester. They returned in 1831 with 2,471 barrels of sperm whale oil.[43]

Mathias Burns of Portland shipped aboard *Jones* under master Francis Davis for its 1826 voyage to South America and returned with 1,827 barrels

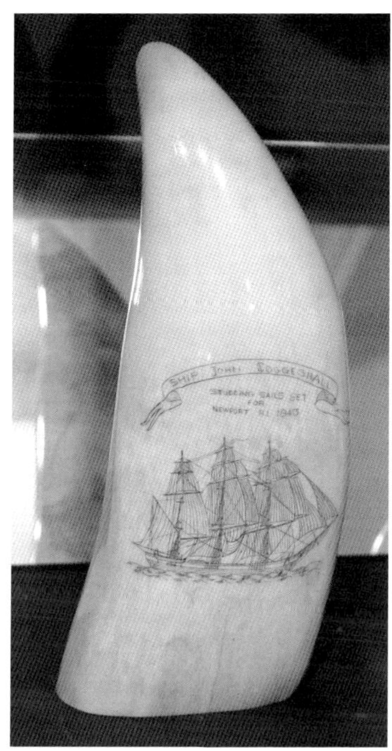

Penobscot Marine Museum scrimshaw sperm whale tooth of whaler *John Coggeshall*. *Photo by author.*

of oil, 140 of them sperm whale. Mainer E.D. Judan shipped aboard *Jones* in 1841 under master William Sisson. The crew returned with 1,340 barrels of oil, 140 of which were sperm whale. *Jones* was broken up at New London in 1842. Logbooks for 1831 and 1832 voyages are in Penobscot Marine Museum in Searsport, Maine. They indicate that three Mainers shipped aboard *Caledonia* for its 1837 voyage under master Benjamin Franklin Pendleton and returned with 1,900 barrels of oil, 250 of them sperm whale. They were David Cutts of Portland, thirteen-year-old William James and William Mans. Another Mainer, George Rodgers, had sailed aboard *Caledonia* in 1835 under John P. Hall. That crew returned from the Falkland Islands with 3,600 barrels of oil, 200 of them sperm whale. *Caledonia*'s logbook for its final 1848 voyage to Patagonia is also in the Penobscot Marine Museum. The museum also holds a water-stained logbook from the whaleship *United States*. Mainer Richard Carter of Scarborough shipped aboard *United States* for its 1843 whaling voyage under Peleg Gifford and returned with 1,150 barrels of sperm whale oil. Samuel Jones of Port Deposit, Maine, was part of its 1856 crew to Indian and Pacific Oceans under Warren Woodward. *United States* was wrecked and abandoned at sea on May 1, 1860. It is not clear if Jones survived.[44]

In 1842, Charles Stephen of Gorham shipped aboard *Charles Phelps* under Captain Palmer Hall. From Stonington, Connecticut, the ship went to the northwest coast and returned March 30, 1844. Thirty-four whales were captured, thirty-three were taken and one was found dead; five were sperm whales, and the remainder were right whales. In the process, thirty-four irons were lost. Ten whales that were killed sank before they could process them. Their lines proved cheap and parted on six whales. Twelve beasts were struck with irons that drew out, as *Charles Phelps* did not carry more expensive toggle-iron harpoons. Yet these challenges were considered average. The crew brought back 2,740 barrels of oil, 140 of which were

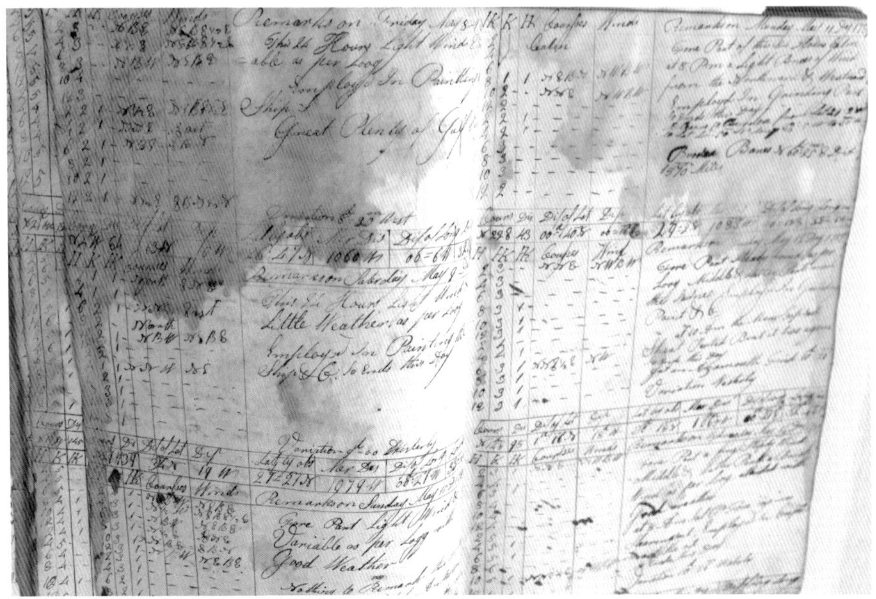

Penobscot Marine Museum's water damaged logbook of whaler *United States*. *Photo by author.*

sperm whale, and 2,600 pounds of baleen. Oil priced at $11.00 per barrel, and they made $28,600. Sperm whale oil at $29.00 per barrel fetched them $4,060.00. Baleen sold for $00.33½ cents per pound, and netted $8,710.00, for a total gross of $41,370.00. Captain Hall was paid $2,544.59. Stephen and rest of crew averaged between $56.23 and $125.12, while boat-steerers, harpooners and mates earned more. Total paid out was $13,289.77, which meant owners took in $28,120.33—over double crew wages. Some crew did not break even, as they had bought clothing and goods from "slop chest" carried and stocked by the captain for sale to the crew. Many had also taken advances before entering *Charles Phelps*.[45]

*Pioneer*'s 1864 voyage showed excellent profits. Under Ebenezer Morgan, it sailed from New London on June 4, 1864. Three Mainers, part of its 1862 crew, were likely still aboard, including Francis Brock of Lebanon, Daniel Moody of Dixmont and Charles Smith of Oxford. *Pioneer* returned September 18, 1865, with 1,391 barrels of oil and over eleven tons of baleen, which sold for $151,060. The cost of outfitting the ship was $35,800, so the percentage of money available to owners, officers and crew was significant.[46]

Besides challenging pay, dangers also included mutiny. In July 1857, *Junior* sailed from New Bedford under Archibald Mellon Jr. with Mainers seventeen-year-old Charles Fifield of Fryeburg and J.W. Hutchins of New Portland. On Christmas Day, they were at S 37°58', E 166°57' in difficult weather. At sundown, they shortened sail, Mellen provided each a glass of grog and most turned in. The next morning, amid a gale, five crew led by a man named Plummer armed themselves with loaded weapons. Five others guarded entry to the forecastle to keep anyone from coming on deck. Captain Mellen was killed as he lay sleeping. The first and second mates were injured, the third mate was killed and mutineers were in control on deck. The second mate was put in irons, the first mate was trapped in the hold and the *Junior* was headed toward Australia. On landfall at Cape Howe, mutineers took two whaleboats and as much plunder as they could. Eight of the ten were caught, including Plummer, and were put on trial. The jury believed their defense of ill-treatment and harsh conditions by Mellen and found only Plummer guilty. Four were cleared of charges, and three were found guilty of manslaughter. Plummer was sentenced to death, but President James Buchanan commuted it to life in prison, where Plummer died. It is not clear what role Fifield or Hutchins played.[47]

Sometimes whalers were tempted by lucrative profits of slave trading. In September 1857, Mainers A.W. Burns and Charles Gower, both from Augusta, shipped aboard *Margaret Scott*. Under Oliver S. Cleaveland, they sailed from New Bedford to the Atlantic and Indian Oceans. *Margaret Scott* belonged to William Tate of New London, who diverted it to the slave trade. It was seized by U.S. marshals, and Cleaveland and Tate were both found guilty of slave trading. *Margaret Scott* was sold at auction and used in the Stone Fleet during the Civil War.[48]

At times, Mainers joined a ship in groups. In 1841, seven signed aboard *William Hamilton* under David Porter West. John Dalton of Whitefield, James B. Goodwin of Stetson, William H. Newlin of Clinton (who left ship in New Ireland), Elijah Norton of Norridgewock as seaman, Thomas J. Sharpleigh of Elliot and Richard Sinceston of Newfield all shipped together from New Bedford for north Pacific waters. On January 27, 1856, with only seventy-five barrels of sperm whale oil, *William Hamilton* was lost off the Chilean coast. It is not clear if anyone survived. In 1841, four Mainers shipped together aboard *William Henry*: James Jones and William Lord, both of Portland; William Wygant of Lubec; and Charles Thompson of China. Under master Godfrey Rider, they departed Provincetown in March 1841 and returned in September with 160 barrels of sperm whale oil.[49]

Several Mainers shipped aboard *Addison* on its four-year voyage from 1856 to 1860. Three hailed from Calais—boat-steerer William F. Heughan and green hands Charles W. Brown and James McElwee. While Heughan returned with *Addison*, the green hands deserted together in November 1857. Francis Finley of Lewiston signed on as green hand, but on December 28, he drowned when his whale boat capsized during a chase. Mainer Hiram John Neal of Vienna signed aboard as seaman in November 1859. He returned and earned notation next to his name as "good boy." That four-year voyage under Samuel Lawrence returned in June 1860 with 2,440 barrels of whale oil, 60 of them sperm whale. *Addison* went into freighting and was lost at Fayal in 1875. Mainers shipped aboard *Florida II* on its 1858–61 voyage out of Fairhaven. Under Thomas W. Williams, Henry McFadden of Wiscasset, Alanson Wardwell of Green and James S. Wyman of Flagstaff all signed as green hands for 1/200 lay each, while James D.F. Dennett of Hollis signed as steward for 1/135 lay. Both Wardwell and Dennett subsequently deserted. *Florida II* returned in October 1861 with 1,410 barrels of whale oil, 750 of them sperm whale. It was abandoned near Point Belcher, Alaska, during the 1871 Arctic disaster.[50]

Some whalers left records of their presence around the world. In 1842, among Aboriginal petroglyphs on islands of Dampier Archipelago off the Pilbara Coast of western Australia, one crewman from whaler *Connecticut* carved an inscription. It was recently discovered on Rosemary Island by archaeologists and local rangers. Dated August 18, 1842, it was carved by eighteen-year-old African American Jacob Anderson and notes that *Connecticut* sailed from New London in August 1841 and had been at sea for a year. Mainer nineteen-year-old Daniel D. Fisher from Falmouth was Anderson's crewmate. Under master Daniel Crocker, they returned in June 1843 with 1,800 barrels of oil, 200 of them sperm whale, and over six tons of baleen.[51]

Three Mainers shipped aboard *Atlantic*, the largest whaling vessel of its day. New London's *Merchant's Magazine* boasted that it was 699 tons and was built in New York in 1836. Its 1845 crew included Henry Beattie and Saul Williams, both of Brunswick, and John Stokes of Bangor. The crew lost master William Beck during the voyage but returned with 5,550 barrels of oil, 50 of them sperm whale, and nearly twenty-five tons of baleen, a cargo worth $80,000. In 1872, Brunswick native John Hinland was aboard *Java*, one of the first whalers to carry a donkey engine on deck for all hoisting work, especially when cutting in and processing. *Java* was under three different masters that season, Edmund Kelley, Eliel T. Fish and Herbert Colson. The ship returned with 3,720 barrels of oil, 320 of which were sperm whale, and over sixteen tons of baleen.[52]

Whaling impacted midcoastal Maine and the Penobscot River and bay area. Many young men answered the call. For instance, 16 participants noted their home as Belfast, Maine. Twelve others hailed from nearby towns like Bucksport, 16 from Rockland, 3 from Hancock, 4 from Islesboro, 2 from Vinalhaven, a dozen from Thomaston, 3 from Camden, 4 from Castine, 3 from Veazie and 8 from Frankfort. Several towns sent 1 or 2 of their young men whaling, such as Appleton, Brewer, Deer Isle, Ellsworth, Hampden, Knox, Liberty, Orono, Freedom, Orrington, Prospect, Searsmont, Stillwater and Union. Bangor sent 138.[53]

For Belfast, the average age was twenty-three years. Thomas Jefferson Burgess, born in 1830, shipped aboard *Franklin*, although it was neither of the Maine-built *Franklin* whalers. His ship was built in Swansea, Massachusetts, in 1832. Burgess's September 1857 voyage, under Johns S. Howland, collected 996 barrels of sperm whale oil before the ship was condemned and sold at Valparaiso on February 15, 1861. Burgess likely suffered an injury while whaling. In 1863, as a thirty-year-old owner and printer of *Progressive Age* newspaper in Belfast, his name was drawn for a draft notice. He reported to the surgeon for his required physical and was found unfit for military duty. It was determined after he was "carefully examined, and is found to be unfit for military duty by reason of Injured Trochanter Major, and, in consequences thereof, he is exempt from service under the present draft." Other Belfast whaling men included twenty-two-year-old William Jones, who shipped aboard schooner *Armadillo* in 1866. It was lost with all hands off St. Eustacia in 1867. Twenty-five-year-old Moses Ewell was on his second whaling voyage in 1859 when his bark *Newark* was lost off Malay Archipelago on April 7, 1863. In June 1855, seventeen-year-old Charles H. Williams shipped aboard *Golconda* out of New Bedford under Philip Howland and returned in June 1859 with 1,692 barrels of oil, all but 120 of them sperm whale. Twenty-year-old Rufus A. Whittier shipped on bark *Nye* in 1860 and was aboard when it was captured and burned by Confederate raider *Alabama* in 1863. Isaac Heal was aboard whaler *Martha II* when it, too, was captured and burned by a Confederate raider, this one CSS *Shenandoah* on June 28, 1865.[54]

For Mainers definitively identified in database, only a few appear to have still been involved in whaling by the early 1900s. Camden's twenty-two-year-old Victor C. Howell departed New Bedford aboard schooner *Ellen A. Swift* on April 28, 1905, under Antone J. Mandly and returned five months later with 450 barrels of sperm whale oil. On July 15, 1907, schooner *John R. Manta* left Provincetown with another Camden native, twenty-two-year-old

Russell Aley, as well as Arthur W. Maxfield of Portland. They sailed under Jose Luis, also recorded as Joseph Lewis, and returned September 1908 with 550 barrels of sperm whale oil. *John R. Manta* was New Bedford's last whaler to be sent out in 1927. Another Mainer was thirty-year-old James L.G. Wright of Portland, who departed New Bedford aboard bark *Morning Star* on May 14, 1910, under Valentine Rosa and returned in August 1912 with a staggering 3,050 barrels of sperm whale oil. Mainer Frank W. Proctor of Somerset had signed aboard the ship in 1908 under masters Benjamin A. Higgins and Frank E. Taber. They returned with 2,700 barrels of oil, 2,450 of which were sperm whale.[55]

Belfast resident Thomas Jefferson Burgess. *Belfast Historical Society and Museum.*

Nantucket whaling captain Benjamin C. Gardner was born in Maine and made two voyages around Cape Horn. Louise Sturtevant, his wife, accompanied him. In January 1826, he sailed *Martha* to Pacific waters and returned in April 1828 with 1,843 barrels of sperm whale oil. *Martha* was built in Pembroke, Massachusetts, in 1810. It was broken up at San Francisco. In August 1845, Gardner took 350-ton *Nantucket* to the Pacific and on January 8, 1847, made a port call at the Pitcairn Islands. It was recorded that *Nantucket* was almost seventeen months out of Nantucket with 800 barrels of sperm whale oil. When the ship stopped at Paita, Peru, in 1848, Gardner agreed to bring home a woman, Ann Johnson, who had posed as a sailor aboard whaler *Christopher Mitchell*. When found out, she was deposited with the American consul at Paita. Gardner was able to bring Johnson back to the United States with propriety because his wife was aboard. *Nantucket* landed at Woods Hole, Massachusetts, where Johnson was deposited in January 1850. The crew also brought 2,051 barrels of sperm whale oil. *Nantucket*, built in 1837, was lost at Nashawena Island in Buzzard's Bay in 1859.[56]

Edward Merrill was born in Durham, Maine, on July 14, 1800. Son of Roger and Dorothy (Cushing) Merrill, he was two when they moved to Portland and eleven when he ran away to sea. Merrill made two voyages out of New Bedford as master aboard Maine-built *Condor* in 1829–30 and 1830–31. He first returned with 2,137 barrels of oil, 267 of which were sperm whale, and seven tons of baleen. The *Condor*'s next voyage returned

Captain Edward Merrill as a young man. *LaFrance Hospitality, New Bedford, Massachusetts.*

with 2,800 barrels of oil, 170 of which were sperm whale. Next, Merrill was master of Maine-built bark *Clarice* for its May 1835 voyage to Atlantic waters. The crew returned in August 1836 with 1,086 barrels of oil, 474 of them sperm whale. Merrill married Mary Converse, relocated to New Bedford and became an inventor, merchant and manufacturer. He refined whale oil and manufactured candles and owned a whale ship and a wharf. On March 28, 1838, Merrill's design for a hydrostatic press for whale oil was patented and cost half as much to manufacture as competing presses. His estate included a New Bedford mansion, land and livestock on Nashaweena Island in Buzzards Bay and land in Florence, California. Today, his wharf is part of New Bedford's historic district and includes a stone countinghouse built by Merrill. The wharf functions pretty much as it did back then as a working pier for the New Bedford fishing fleet. He also built Homer's Wharf. Edward Merrill died on September 11, 1884, and was buried at Rural Cemetery in New Bedford.[57]

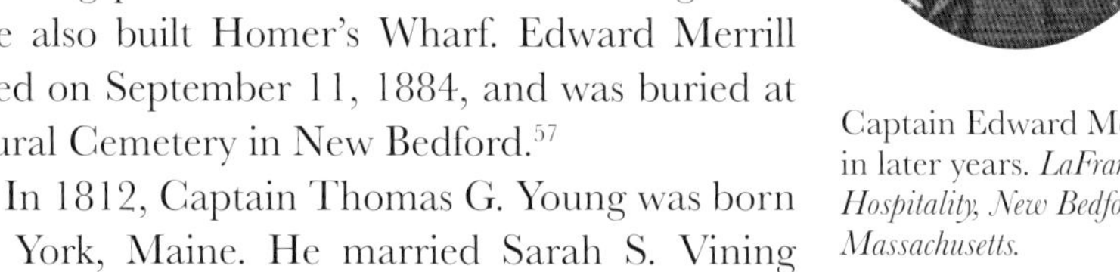

Captain Edward Merrill in later years. *LaFrance Hospitality, New Bedford, Massachusetts.*

In 1812, Captain Thomas G. Young was born in York, Maine. He married Sarah S. Vining and made five voyages, mostly out of Fairhaven aboard whaling bark *Favorite*. His November 1840 voyage to the Pacific included Mainers Henry F. Hussey of Vassalboro, Lewis Jones of Turner, Thomas T. Houlton of Hollis and David Lewis of Boothbay. They returned in June 1843 with 1,848 barrels of oil, 848 of them sperm whale, and 4 tons of baleen. Young's August 1846 voyage included Isaac Mcleod of Bristol and returned in November 1849 with 1,800 barrels of oil, all but 250 sperm whale, and 1,400 pounds of baleen. Young was master of *St. Peter* in 1852 out of New Bedford and collected 788 barrels of oil, 379 of which were sperm whale, and over 32 tons of baleen. *St. Peter* was wrecked by its crew off Chatham Island, New Zealand, on March 24, 1855. Young survived and returned to Fairhaven. He sailed *Favorite* again to the north Pacific in 1863, but the ship was burned by Confederate raider *Shenandoah* on June 28, 1865. Losing two ships seemed to be enough for Young. He retired and died in Maine on November 1, 1875.[58]

In 1818, Captain Gersham Leander Cox was born in Maine and was master for five whaling voyages from New Bedford. On *Brighton*, he went to the Indian Ocean in August 1842 and returned in July 1844 with 2,543 barrels of oil, 116 of which were sperm whale, and twelve tons of baleen. Cox's October 1844 voyage included Mainers Joseph C. Coleman and George G. Wells, both of Hallowell; George F. Jamerson of Windsor; and Athul Ramsdell of Lubec. They returned in April 1847 from the northwest coast of America with 2,612 barrels of oil, 148 of them sperm whale, and nearly fifteen tons of baleen. Cox next sailed *Magnolia* to the Pacific twice. Its September 1851 voyage returned with 3,909 barrels of oil, 294 of them sperm whale, and nearly twenty-five tons of baleen. Cox's October 1854 to May 1858 voyage returned with 2,572 barrels of oil, 56 of them sperm whale and over sixteen tons of baleen.[59]

A relation of Cox was James Valentine Cox, born in Hallowell, Maine, on July 1, 1818. He made two voyages to the Pacific as master on *Abigail* out of New Bedford and returned with a combined 2,000 barrels of sperm whale oil. His other two voyages were aboard bark *Draco*. Its December 1843 voyage returned in April 1847 with 1,631 barrels of sperm whale oil. Cox sailed *Draco* again in August 1847 and returned November 1850 with 1,382 barrels of sperm whale oil. He died on November 23, 1884.[60]

Captain George A. Smith was born in Maine in 1843 and made four voyages, mostly out of New Bedford, two as master of *Charles W. Morgan*. Its October 1886 voyage to the north Pacific returned in November 1887 with 1,325 barrels of oil, 275 of which were sperm whale, and over five tons of baleen. In December, Smith took *Charles W. Morgan* out of San Francisco to Japan and the Okhotsk Sea and returned in November 1888 with 560 barrels of oil, 135 of them sperm whale, and 4,500 pounds of baleen. He next took bark *Triton* to the north Pacific in November 1888 and returned in November 1889 with no record of any catch. Smith was master of bark *Sunbeam* in August 1890 to Atlantic waters and returned in June 1893 with 1,150 barrels of sperm whale oil.[61]

Another Maine captain was East Limington native Amos Chase. He was born in 1828. As a young man, Chase served aboard whalers and worked his way to captain in eleven years. As captain, he sailed *Hector* from New Bedford to the Pacific from November 1856 to July 1860 and returned with 1,804 barrels of sperm whale oil, which earned him 1/14 lay. William M. Ray, twenty-two-year-old Bath resident was part of its crew. Chase sailed *Hector* in 1861, and the crew included twenty-one-year-old Mainer Alvin S. Warren of Lovell. By April 1865, the crew had collected 400 barrels of oil, 260 of them sperm whale, as well as 1,850 pounds of baleen. *Hector* was burned by Confederate raider CSS *Shenandoah* at Lodh Pah harbor on April 4, 1865. Chase next did two back-to-back four-year voyages aboard bark *Platina* out of Westport and then two more out of New Bedford with *Jireh Perry*. His final set of two back-to-back four-year voyages was aboard New Bedford's *Niger*. He returned in July 1890 after thirty-four years of near-constant whaling.

Gravestone of Amos Chase. *Photo by John F. Berube.*

Amos Chase died in 1915 in Westbrook, Maine, and was buried there at Saccarappa Cemetery.[62]

Another prolific captain from Maine was Leander C. Owen, a native of Brownfield. He moved to New Bedford as a young man and went into whaling at age fifteen. He remained in the profession for the next forty years. His career was interrupted by war—he served in the Union navy under Admiral David Porter during the capture of Fort Fisher. Owen suffered hearing loss as a result and spent the rest of his life seeking compensation from the government. In 1871, his first whaling voyage as master was aboard *Contest* to the Bering Sea, where she passed Indian Point on June 6 and took aboard nine men from stranded whale ship *Japan*. The Bering Strait was heavily blocked with ice, but *Contest* still caught two bowhead whales, which produced two hundred barrels of oil. By June 16, the ship had drifted past East Cape and came upon walruses and took four hundred over the next twenty-three days, which rendered another three hundred barrels. During the night of June 17, *Contest* got into ice in heavy fog and sustained bow damage before working its way out. With damages repaired, Owen worked farther north past Icy Cape. By August 28, after taking five bowheads, *Contest* went aground off Wainwright Inlet and had to be assisted by two nearby whalers. By September 7, Owen and other captains had sounded waters hoping for a way out of this tightening noose of ice. They found less than nine feet of water off Wainwright Inlet. *Contest* was beset, and Owen was forced to abandon it and send his crew on their way to safety by whaleboats. He was one of several captains who signed a general letter detailing their plight and decision to abandon ships and catches, as well as a follow-up document asking for help. The loss of *Contest* was valued at $40,000.[63]

Owen next took New Bedford's *Jireh Perry* to Arctic waters from December 1871 to April 1875 and returned with 4,550 barrels of oil and thirty-six tons of baleen. There, he spotted *Helen Snow* abandoned deep in an icepack. Owen managed to get a salvage crew aboard the ship, work it free of ice and get back to San Francisco. It had 209 barrels of oil aboard, 169 of them sperm whale oil. Once in San Francisco, *Helen Snow* was put up for sale. Bought by Russians, it was renamed *Tugar* and later *Desmond.* Owen's ice experience convinced him that to increase catches, he needed ships equipped with auxiliary power: "It seems an open season and I have no doubt but a steamer could have gone to Mackenzie River, and I don't know but clear around the northwest passage." In 1875, he sailed *Three Brothers* out of New Bedford and bark *Coral* in 1878, which netted 6,063 barrels of oil and nearly twenty-six tons of baleen. Owen next sailed the first steam bark—Bath-built *Mary and Helen*—to Arctic waters from September 1879 to October 1880. Aboard were four Mainers, James Brown of Portland, Henry R. Farrin of Bath, Frank J. Smith of Bangor and Winfield Trott of Eastport. They arrived in San Francisco with 2,350 barrels of oil and almost twenty-three tons of baleen from twenty-seven bowhead whales—one of the largest catches in Arctic fishery history, valued at over $100,000. *Mary and Helen*'s investors recouped their money after just one season.[64]

Owen next sailed Maine-built steamer *Belvedere*, which included Bath residents Herbert C. Shea and O.W. Stacy. *Belvedere* hauled 1,800 barrels of oil and seventeen tons of baleen and rescued survivors of wrecked whaler *Daniel Webster*. Then, in 1882, Owen sailed Maine-built *North Star*, but there is no record of any catch. In the *North Star*, Owen did rescue the crew of USS *Rodgers*, formerly *Mary and Helen*, which had been sent north to search for *Jeannette* expedition survivors. *Rodgers* had burned to its waterline that winter. In 1882, the *North Star* put in to North Head, Saint Lawrence Bay, where natives complained of lack of walruses due to prolific harvesting from

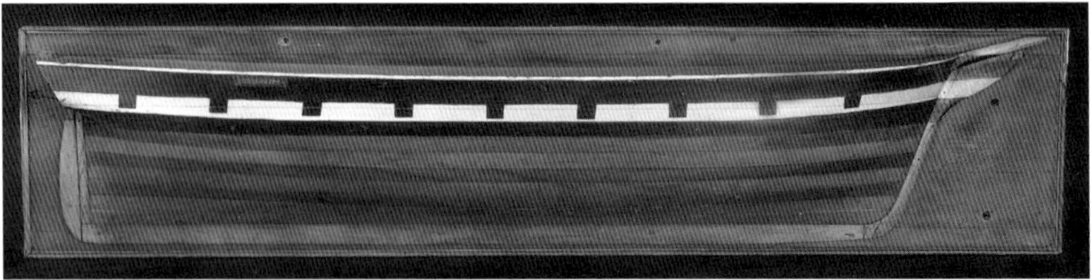

Half-hull model of *Jireh Swift* sailed by Captain Amos Chase. *Division of Work and Industry, National Museum of American History, Smithsonian Institution.*

experts such as Owen. It reached Sea Horse Islands eighty miles southeast of Point Barrow on June 25, 1882, which was the earliest date reached by any vessel. On July 8, *North Star* was crushed in ice two miles from shore, and its crew made its way to a nearby U.S. Army weather station. Owen took bark *Rainbow* to the north Pacific, where he had to deal with a murder aboard ship. In 1884, he sailed Maine-built *Thrasher* as a new hire for the recently founded Pacific Steam Whaling Company. *Thrasher* sported steam tryworks and iron oil tanks. Owen returned with 1,800 barrels and nearly twenty-three tons of baleen. *Thrasher* was sold to Seattle owners, renamed *Kamchatka* and was crushed in ice. Owen next made three voyages with *Susan and Mary*. On the first one, he struck and killed only six whales but found and killed 250 walruses on one piece of ice. During his third voyage in 1887, Owen was wrecked in an Arctic gale and was rescued by revenue cutter *Bear*. Owen then sailed bark *Sea Ranger* from December 1889 to November 1890 and brought back 290 barrels of oil and seven thousand pounds of baleen. He then sailed from San Francisco and brought *Susan and Mary* back around Cape Horn to New Bedford after its 1886 whaling season and returned with 750 barrels of whale oil and seventeen thousand pounds of baleen. Captain Leander C. Owen died on March 25, 1911, age seventy-seven, and was buried at Oak Grove Cemetery in Tisbury, Massachusetts.[65]

Some captains retired to Maine after their years at sea. One was Christopher Pinkham, born in 1730, who made three voyages out of Nantucket. In 1764, he was master of the schooner rigged sloop *Dolphin*, although there was no record of any catch. There are also no results from 1767 in a schooner named *Falmouth*. Pinkham's third voyage was in 1775, aboard an unknown vessel. He married Ruth Folger and died somewhere in Maine, although details are not clear. Another captain out of Nantucket, Benjamin Worth, was born on October 11, 1748, and died in East Vassalboro, Maine, on August 25, 1838. In 1775, he took *Hector* whaling, and there are no records of any catch. He sailed *Hector* again in 1777 with no record of any catch. Worth gave up whaling, became a preacher and relocated to Maine. His wife was Phoebe Coffin.[66]

Obed Paddock was another Nantucket-born captain who retired to Maine. Born on November 6, 1762, he made four voyages aboard *Falkland* starting in 1787 with no record of any catch. He sailed *Olive Branch* twice in 1795–96 and 1797–99, the second voyage rounded Cape Horn and reached Chile and returned with 1,100 barrels of oil, 1,000 of them sperm whale. He sailed *Fortitude* to Delagoa Bay in present-day Mozambique and returned in March 1799 with 1,000 barrels of oil. Paddock married Margaret Macy. He retired to Fairfield, Maine, in 1800. In failing health, Paddock was housebound for his last thirteen years. A devout Quaker, he died at age seventy-two on January 10, 1834.[67]

Charles F. Coffin, another Nantucket-born captain who retired to Maine, was born on April 18, 1794. He married Mary Folsom and made one whaling voyage from Nantucket in 1836 as master of *Washington*. The ship rounded Cape Horn, but Coffin left ship at some point and was replaced by Stephen Bailey. The crew returned in December 1839 with 1,780 barrels of sperm whale oil. Coffin died in Maine on December 19, 1873.[68]

Massachusetts native Captain Edmund Crowell was born in 1807. He married Jane W. Grafton and made one voyage to the Pacific as master of *Clifford Wayne*. Its March 1841 voyage out of Fairhaven was marked by a mini mutiny. On Tuesday, January 9, 1844, when Crowell ordered a man

Stereo view of Merrill's Wharf in New Bedford, Massachusetts. *New Bedford Free Public Library.*

Modern-day view of Merrill's Wharf looking toward shore. *Photo by author.*

put in irons for insolence and disobeying orders, the rest of the crew refused to do duty. There is no information on how such impasse was broken, but the transgressor was released, and the crew returned to duty. They arrived in July 1844 with 1,400 barrels of sperm whale oil. Crowell retired and died in Union, Maine, on January 26, 1886.[69]

James Harvey Shearman Jr. was a New Bedford–born captain who became a farmer in Sidney, Maine. He sailed twice as master aboard *Young Phoenix*, starting in 1836. It first returned in March 1840 with 2,397 barrels of sperm whale oil. He then sailed *Coral* in 1839 and returned in 1842 with 3,118 barrels of sperm whale oil. Back to *Young Phenix* to finish out her voyage, Shearman returned in August 1844 with 2,750 barrels of sperm whale oil. He then took *Emigrant* to the Indian Ocean twice out of Bristol, Rhode Island. Its February 1843 voyage returned in September 1844 with 500 barrels of oil and a ton of baleen. Shearman's last voyage in November 1844 was on *Emigrant*, and he returned in February 1847 with 402 barrels of oil, 272 of them sperm whale. He became a merchant captain, participated in the California gold rush and retired to Sidney, where he took up farming and died in 1890.[70]

*Lively*, 102-ton, clipper-built and schooner-rigged, was launched in 1851 in Chelsea, Maine, near Gardiner. It was sixty-eight by twenty-two by eight feet and, in July 1851, went on its maiden whaling voyage out of Fairhaven. Aboard were forty-two-year-old Hallowell native Job Pierce as the managing owner/builder and John Johnson as captain. By 1852, it was reported lost with all hands, including three Mainers from Hallowell, one from Vassalboro and one from Eastport.[71] According to William A. Baker's account, "Those that stayed with the schooner until 10 October 1852 went to their deaths. On that day the bark *Anna of Bremen* sent a boat to investigate a capsized schooner off Bermuda in a region swept by heavy gales two days earlier. The wreck was that of the Chelsea-built whaler."[72]

Another Hallowell whaling connection was Job Pierce's brother, Ebenezer. Born 1817, he was seven years behind brother Job but followed him to the sea. He was the inventor/creator of a bomb lance harpoon, considered the deadliest whaling weapon ever used. In 1878, New Bedford gun manufacturers Pierce and Selmar Eggers obtained a patent that focused on a means of locking the breech block in place for discharge on a breech-loaded shoulder gun (U.S. patent no. 200,338, February 12, 1878). The breech block, with firing mechanism, trigger and trigger guard, was hinged at the bottom so it could open by pivoting down and forward. On opening, a center-fire cartridge was inserted, and the bomb lance was loaded down its muzzle. Breech block was then closed and locked by a feature patented by Pierce.[73] An improved all-brass breech-loaded shoulder gun was then patented by Pierce in 1882 (U.S. Patent no. 255,330, March 21, 1882). The barrel opened by tilting forward to load the bomb lance and cartridge into its breech. The spring-loaded locking plate secured it in closed position for firing. On each side, near its breech, was a large lug cast as part of the barrel. When locked, lugs entered recesses along its brass stock.

Pierce explained:

> *During the act of firing the shock and strain will be transmitted from the barrel to its lugs, and hence the walls of the recesses within which the lugs are fitted will receive the impact of recoil. In this class of guns the recoil is very heavy, and hence great strength of parts is essential. These lugs being arranged on the sides of the barrel directly opposite each other, and about in line with the axis of the barrel or its longitudinal center, and being received into the said recesses, will entirely relieve the pivot by which the barrel is connected with the stock from all strain during firing, and will also relieve the locking plate or latch from pressure and strain; also, they will so hold the barrel that but a slight locking device will be required for maintaining the rear end of the barrel in position against the breech-piece.*[74]

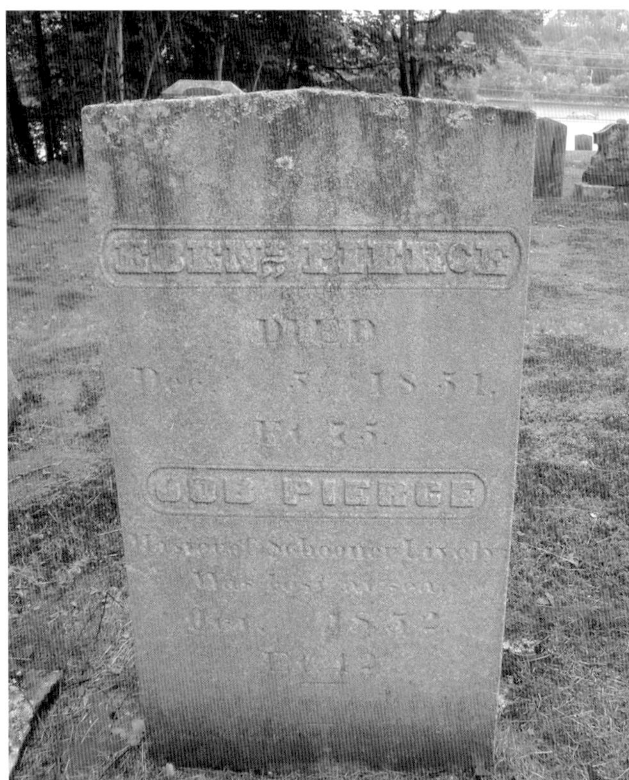


*Right*: Gravestone of Eben Pierce and Job Pierce in Farmingdale, Maine. *Photo by author.*

*Below*: Eben Pierce's shoulder-fired bomb lance gun. *Division of Work and Industry, National Museum of American History, Smithsonian Institution.*

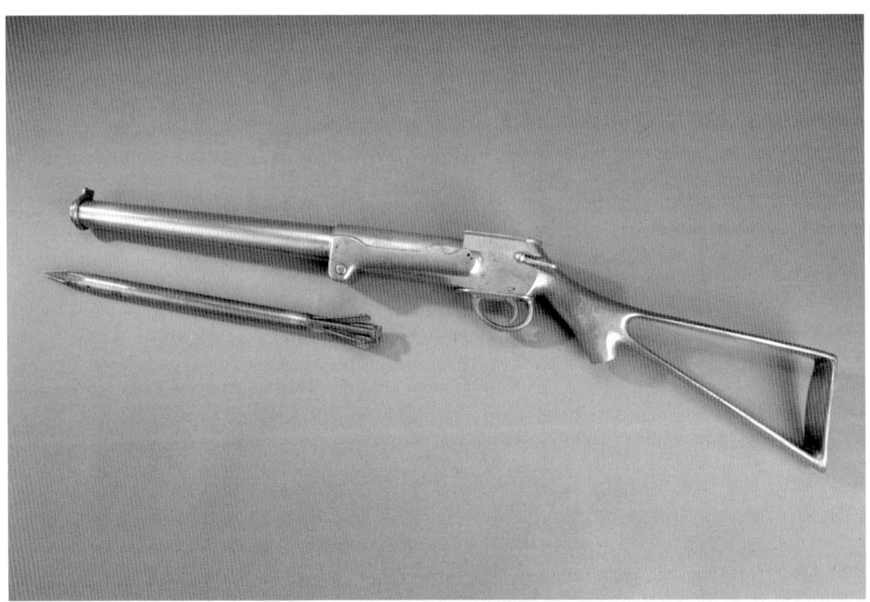

Memorial Stone of Eben Pierce. *Photo by author.*

His gun was intended to fire a bomb lance with metal flip-out fins for stabilization in flight by means of a cartridge. In storage, fins were held in place by a sliding iron ring. Pierce's bomb lance was patented in 1879 (U.S. Patent no. 211,778, January 28, 1879).[75] Captain Eben Pierce died in 1902 at the age of eighty-five and is buried near his brother, Job, at Maine Avenue Cemetery in Farmingdale, Maine.

Another whaling connection was Abel Douglass, born on Isle au Haut in 1841. He traveled with his family to San Francisco by way of Cape Horn in 1849. By the 1860s, Abel and brother, Albert, had migrated to Victoria to start a whaling business with partner James Dawson. Dawson and Douglass Whaling dominated the decade, posting best records for whaling catches. Needing a whaling station, they founded Whaletown in 1869 on Cortes Island just off British Columbia. Today, Whaletown is known as the gateway to Cortes Island and is serviced by a ferry. Douglass served as captain and went into sealing by the 1890s. He was involved in an international incident when one of his ships was seized by American authorities in a conflict between Canada and United States. *May Belle* was taken to an Alaska port and held until conflict resolved, although Douglass did not live to see it. He died in Seattle in 1907 and was buried at Lake View Cemetery in Seattle.[76]

3.

# START OF AN ERA

*The ocean is an object of no small terror.*
*—Edmund Burke*

Most whalers operated from Massachusetts, Connecticut, New York or California. Early on, some ports tried specializing, such as Sag Harbor, which focused on northern right whales. New London and Stonington sought sea elephants and right whales. New Bedford, however, went after profits and took any oil-bearing sea creature it came across. That approach dominated, especially when whale numbers decreased and more ships entered whaling. American whaling peaked in 1846, with 680 ships and barks, 34 brigs, and 22 schooners registered in thirty-four different U.S. ports—one quarter operating in the Atlantic, Hudson Bay and Davis Strait. At its peak, American whaling employed over seventy thousand men and ranged throughout all oceans—a true golden age. Mainers shipped aboard those vessels, over two hundred of which were built in Maine. While not initially intended for whaling, many found their way into service, one source suggests over half. In 1784, Mainers in Georgetown built a schooner named *Betsey*, which made three whaling voyages to north Atlantic waters out of Boston between 1794 and 1800, under Joseph Hatch. There is no record of its catch.[77]

In 1784, Biddeford-built brig *Sea Horse* was launched and made four voyages from 1788 to 1795 from Gloucester to Cape of Good Hope. The crew returned with eight hundred barrels of whale oil and five tons of

Etching of a view of the whale fishery in Greenland. *Library of Congress.*

baleen. *Commerce*, built in Bath in 1795, sailed to the south Atlantic twice in 1797 and 1798, first under Henry Deshon and then Stephen Rawson. Rawson and *Commerce* were captured in 1798. *Rover*, launched in Kennebunk in 1801, made one whaling voyage to Cape of Good Hope from 1803 to 1805 under Isiah Ray and returned in January 1805 with nine hundred barrels of sperm whale oil.[78]

In 1803, Kennebunk-built *Charles* was launched and made three voyages out of New Bedford between 1804 and 1809. On one trip, *Charles* brought back 2,200 barrels of sperm whale oil. In 1804, Sedgwick schooner *Liberty* was launched and made voyages from Fairhaven between 1815 and 1816. Its first trip, under Ebenezer Hathaway, returned in September 1815 with 100 barrels of sperm whale oil. Hathaway took it out again until October but recorded no catch. Under Peter Butler, *Liberty* brought back 35 barrels of sperm whale oil. In 1804, *Mary Ann* was built in Columbia, Maine, and sailed twice to the south Atlantic between 1829 and 1831. Its 1830 voyage to Brazil under Abraham Swain brought back 1,000 barrels of oil. *Mary Ann* wrecked off Gay Head, Martha's Vineyard, in June 1831. *Frederick Augustus*, also Columbia-built, was launched in 1805 and made four three-year voyages from Newport, Rhode Island, to the Pacific from 1821 to 1833. Its first three, under Joseph Earl Jr., resulted in 6,800 barrels of sperm whale oil. Its final voyage, under William Kurn, brought back 1,600 barrels of sperm whale oil. It was broken up in 1834.[79]

In 1809, Wiscasset-built *Columbus* was launched and made eleven voyages between 1831 and 1854 to the south Atlantic, Indian Ocean and north Pacific, as well as New Zealand. *Columbus* returned a total of 38,565 barrels of whale oil, 5,298 of which were sperm whale, and nearly sixty-five tons of baleen. After a half century, *Columbus* was broken up in 1859. A different *Columbus*, not from Maine, made its 1839 voyage from New London to the south Atlantic under Caleb Strong Holt with eighteen-year-old Portland native Henry Morgan. The crew returned with 450 barrels of sperm whale oil.[80]

At least three survivors of *Essex*, whose sperm whale encounter Herman Melville adapted to his classic *Moby-Dick*, went back to sea on a Maine-built whaler. Thomas Nickerson and Charles Ramsdell shipped aboard *Two Brothers* in 1821. Its master was George Pollard Jr., who had lost *Essex*. It was 217 tons, eighty-four feet long and built in Hallowell in 1804. In 1823, two years into its voyage, Pollard's ship foundered in a storm on French Frigate Shoals, and men escaped into whaleboats. The following day, *Martha* came to their assistance. Although everyone survived, Pollard confessed during his return to Nantucket, "no owner will ever trust me with a whaleship again, for all will say I am an unlucky man." In February 2011, archaeologists located *Two Brothers* in waters near Shark Island, an uninhabited islet six hundred miles northwest of Honolulu, Hawaii. They spotted a giant anchor twenty feet down and saw three cast-iron cauldrons called trypots, which were used by whalers to melt blubber into oil. They next found a trail of sunken bricks, *Two Brothers*'s ballast and iron harpoon tips, perfectly matching 1820s Nantucket manufacturing. These helped identify *Two Brothers*, as it was the only whaling ship lost there in that decade. Shards of cooking pots also matched advertisements in Nantucket newspapers.[81]

In 1810, *Eliza Barker* was built in Hanover on Androscoggin River and made two voyages from New York to the western Pacific and Japan. Under David Paddock, it returned in November 1819 with 2,100 barrels of whale oil, 1,950 of them sperm whale. It had been boarded by pirates, who robbed the crew members of their clothing. Obed Alley sailed in December 1820 from Hudson and returned in June 1823 with 1,550 barrels of sperm whale oil. Bowdoinham-built brig *Liberty* made one voyage out of Newport, Rhode Island, to South Africa in 1816. Under Amaziah Gardner, the ship returned in 1817 with 200 barrels of sperm whale oil. Columbia-built *Robinson Potter*, under Reuben Swain, sailed in July 1820 from Newport to the Pacific and returned in December 1822 with 2,100 barrels of sperm whale oil. It was lost in 1825. Bath-built brig *William and Nancy* made four voyages from Nantucket to the south Atlantic. In 1816, it sailed to Guinea under Coffin Whippey and returned with 150 barrels of sperm whale oil. In 1820, Tristam Folger took it out but returned empty. In 1822, it was sold to new owners in Falmouth, Massachusetts, and then in 1828 to owners in Rochester, New York.[82]

In 1810, Thomaston-built *General Knox* was launched and made one voyage under William B. Orne in 1820. It sailed with Salem Fleet, which included *Nancy* and tender *Governor Brooks*. They visited the Falkland Islands and were spotted in Berkeley Sound, South Shetlands, on October 17, 1820, by Nathaniel Palmer of Stonington brig *Hero*. Palmer, credited with the first sighting of Antarctica, again encountered *General Knox* when he anchored beside it in West Point Harbor on October 20. *General Knox* returned in eighty-seven days from the Falkland Islands on June 5, 1821, and brought five thousand fur seal pelts and six hundred barrels of sea elephant oil.[83]

Another 1810 vessel was *Jason*, built in Kittery. Out of New London, it made nine voyages between 1832 and 1848. Its first trip to the Pacific, under William Elias Coite, brought back 2,500 barrels of oil. Its next three, under Luther Fuller, yielded a total of 4,400 barrels of oil and nineteen thousand pounds of baleen. *Jason*'s eighth voyage, under Jeremiah Slate, visited Tristan da Cunha, Kerguelen Island, St. Paul's Islands and Amsterdam Island and included Mainers William Fergissen of Hancock, Nathan Munay of Berwick and William Williams of Bangor. They brought back 2,650 barrels of oil. Its final voyage, under Elijah B. Morgan, accompanied tender *Exile* and visited Kerguelen. The crew returned in 1848 with 2,600 barrels of oil and eight tons of baleen. In 1811, brig *Mary* was built in Scarborough and sailed four times to the south Atlantic, starting in 1815 under Paul Howland. Out of New Bedford, it returned in March 1816 with 1,300 barrels of oil. Its next three voyages came up empty. *Mary* was reported lost near Cape Blaco off the southwest coast of Africa on May 28, 1818.[84]

In 1814, Orrington built 285-ton *George Porter*. It made fourteen voyages between 1819 and 1850, many from Nantucket to Brazilian waters. In 1819, it returned with 760 barrels of oil, seventy-six of them sperm whale. From 1821 to 1822, it sailed under Prince B. Moores to South Shetland Islands in company with schooner *Harmony*, under a master Hodges. *Harmony* was lost in the Indian Ocean on February 8, 1839. Several *George Porter* voyages included Mainers such as Eastport resident Samuel Smith, who did two voyages in 1828 and 1829. Both voyages totaled over 1,900 barrels of oil and twelve tons of baleen. Mainer Peter Beeden accompanied *George Porter* in 1832. They returned with 1,850 barrels of oil and fifteen thousand pounds of baleen. Mainer James B. Hall shipped in 1833 and brought back 1,703 barrels of oil. Aboard was a Machias resident named Henry. On its last four voyages, *George Porter* visited the south Atlantic, Pacific and Indian Oceans and brought in over 4,000 barrels of oil. Three Mainers from Sidney shipped in 1834: Joshua Davis Jr., William Harding Jr. and Jothand A. Sawtill. Three Mainers sailed in 1848: Thomas Burgess of Castine, Nathaniel Hooper of Sanford and Ingraham D. Oliver of Bath. *George Porter* was wrecked and condemned at Mahe in Seychelles in 1850. Its 700 barrels of oil had to be shipped home.[85]

In 1815, Portland-built *Friendship* was launched. It sailed to the Pacific and Indian Oceans, New Zealand waters and America's northwest coast starting in 1831. Its 1844 voyage included Mainers Henry F. Briggs of Bradford and Moses W. Stockman of Danville and returned in February 1846 with 2,750 barrels of oil, 350 of which were sperm whale, and nearly thirteen tons of baleen. Kennebunk-built schooner *Packet* made two voyages out of Fairhaven to south Atlantic waters in 1822 and 1823 under Thomas Delano Jr. and brought back 610 barrels of sperm whale oil.[86]

In 1818, Portland-built *Bourbon* was launched. It took two voyages to the south Atlantic between 1821 and 1824 under Latham Paddack. The second trip brought back 1,600 barrels of oil. *Bourbon* was broken up in 1843. Falmouth-built *Liverpool* made fifteen whaling voyages to the south Atlantic, north Pacific and Indian Oceans between 1831 and 1860. Its June 1840 voyage out of New Bedford, under Joseph Thomas, included Peter L. West of Freeport. The crew returned with 2,528 barrels of oil, 263 of them sperm whale. Its 1858 voyage under Charles D. Davenport included Chester Sprague of Appleton and returned in 1860 with 90 barrels of sperm whale oil. *Liverpool* was sold to New York owners in 1860.[87]

Brewer built the brig *Levant* in 1819. It made one voyage in 1835 out of Lynn, Massachusetts, under William Caldwell. No catch was recorded. Kennebunk-built *Maine* made eleven voyages to the south Atlantic, the Indian Ocean and America's northwest coast, starting in 1829. Its 1840 voyage out of Fairhaven under James Magee included Mainer William D. Stone of Shapleigh, and returned in 1842 with 1722 barrels of oil, 266 of them sperm whale. *Maine*'s final 1846 voyage, under George E. Netcher, included Andrew J. Dolloff of Rumford and had accumulated 1,400 barrels of oil and 6,900 pounds of baleen when it was lost near Columbia River on August 25, 1848. Orland-built *Sarah Lee* made five voyages from Bristol, Rhode Island, starting in 1835. Under William Weeks, James Darling, William M. Blye (twice) and Nelson Waldron, *Sarah Lee* brought back 1,820 barrels of oil, 80 of them sperm whale. Its last

Maine Maritime Museum scrimshaw collection including a whale bone swift. *Photo by author.*

two voyages returned empty, which prompted the owners to sell it into merchant service. *Sarah Lee* was lost in 1842.[88]

In 1820, the same year Maine became the twenty-third state, Bath-built *Mary* was launched. It made fourteen voyages to the north Pacific, the Indian Ocean and New Zealand waters, starting in 1836. Its 1854 voyage out of New Bedford, under Silas Cottle, included Franklin Littlefield of Solon, Maine. The ship returned in August 1856 with 837 barrels of oil, 668 of them sperm whale, and a half ton of baleen. Its 1856 voyage under John R. Sands included James Kendall of Calais and returned in May 1859 with 1,258 barrels of oil, 344 of them sperm whale, and one ton of baleen. *Mary*'s final voyage in 1863 had two Mainers, Barzillai B. Gatchele of Boothbay and S. Metcalf of Union. The ship was lost in the Okhotsk Sea in 1864. Bath's 222-ton brig *Jasper* was built by John and George F. Patten and was the first of that family's fleet. Extensively rebuilt and refitted into a bark, *Jasper* was sold to New Bedford owners and put into whaling. It made ten voyages to the south Atlantic, south Pacific and Indian Oceans, starting in 1836. Its 1842 voyage out of New Bedford, under Joseph Bennett Jr., included Mainer M.M. Simonton of Vassalboro and returned with 1,147 barrels of oil, 245 of them sperm whale, and over five tons of baleen. *Jasper* was condemned at New Zealand in September 1853.[89]

Twenty-four whaling vessels were built within five years of Maine's statehood. Kennebunk's *General Pike* was launched in 1821 and made nineteen voyages to south Atlantic, Indian and Arctic waters, as well as New Zealand. Its 1843 voyage included ten Mainers, all from Hallowell—Edward Akin, Lunard Blansher, brothers Horatio and Robert Day, Edwin Goodwin, Joseph Graves, possible brothers Albert and Frederick Meady, Elisha Runnels and George Springer. Albert Meady was *General Pike*'s second officer. Meady fell ill and was discharged on December 13, 1845, at Paita, Peru. Maine native Job Pierce was master and returned with 2,600 barrels of oil, 344 of them sperm whale, and eleven tons of baleen. The ship's September 1856 voyage under James M. Russell included Mainer Greenleaf T. Marinet of Scarborough and returned in March 1859 with 2,822 barrels of oil, 113 of them sperm whale, and eleven thousand pounds of baleen. *General Pike* was captured by Confederate raider *Shenandoah* in the Bering Sea but was spared being torched. Due to its age, it was bonded to carry prisoners back to civilization. *General Pike* was condemned and sold in Tahiti in 1868. Hallowell-built schooner *Laurel* made fifteen voyages to the south Atlantic waters, starting in 1831. Under Joseph R. Tabor, Bartlett Mayhew, Hiram Luce, Charles Wood, Jonathan Worth and George A. Smith, it brought back 2,690 barrels of oil, 2,607 of them sperm whale. *Laurel* wrecked at New Smyrna in 1844. Machias-built *General Gates* made one voyage out of Boston under Abimileh Riggs, but no catch was recorded. Bath-built *Grand Turk* launched in 1822. It made thirteen voyages to the south Atlantic, Indian Ocean and New Zealand waters, starting in 1827. Its May 1834 voyage under David H. Bartlett included Mainer Charles Combs of Bath and returned in January 1836 with 2,550 barrels of oil, 150 of them sperm whale. *Grand Turk* was condemned and broken up in New Bedford in 1843.[90]

Bath-built, 278-ton bark *Arab* was launched in 1823 and made fifteen voyages spanning a forty-year career, starting in 1831. Out of Fairhaven, it sailed to Kerguelen Island in November 1835 and returned with 2,760 barrels of oil, 360 of them sperm whale. In 1838, *Arab* netted 2,400 barrels of whale oil. In two voyages, Captain Ezra Wrightington brought back 3,300 barrels of oil and eighteen tons of baleen. Its 1840 voyage included Mainer Warren Allen of New Portland, and it returned with 2,183 barrels of oil, 428 of them sperm whale. In 1857, John R. Cole of Calais and Boardman Richardson of Vassalboro shipped. In 1858, under William Washburn, *Arab* revisited Kerguelen and Heard Islands with schooner *Oxford* as tender and brought back 2,000 barrels of oil and half a ton of baleen. Its 1860 voyage, under Joseph P. Nye, included Mainers Edwin C. Hinckley of New

Portland and Henry J. Pottle of Harrison. Four Mainers shipped aboard in 1861—John H. Marshall of Lewiston, James Mcelivee of Calais, Charles F. Mesem of Limerick and Henry T. Sattle of Harrison. They returned with 278 barrels of sperm whale oil. In 1862, *Arab* was sold to New London owners and was put in at Cape Town on March 27, 1863. One sailor was arrested by the United States consul and charged with murder and culpable homicide committed on Heard Island. *Arab*'s 1864 voyage under Edwin Church returned with over 1,700 barrels of oil. Its final whaling voyage out of New Bedford from 1867 to 1871 included two Portland residents, James Mann and Edward Mohair. In the Arctic, under Frederick P. Cole, it brought back over 2,000 barrels of oil. *Arab* was removed from service and condemned in 1871.[91]

Quite a few 1823 Maine-built vessels went whaling. Frankfort-built *Canova* made one voyage to New Zealand from 1839 to 1841 under Charles W. Saunders and shipped back 2,650 barrels of whale oil. *Canova* was condemned at Rio de Janeiro in September 1841. Scarborough-built bark *Clarice* made fifteen voyages to the Atlantic, Indian and Pacific Oceans starting in 1835 under Portland captain Edward Merrill. Mainer Joshua Sawyer of Durham sailed *Clarice* in December 1841 under Joseph Dexter and returned in August 1845 with 853 barrels of sperm whale oil. John Grant of Frankfort and William Smith of Standish sailed in 1857 under Frederick W. Brown to the Pacific and returned with 779 barrels of oil, 687 of them sperm whale. Smith was still aboard for the 1863 voyage under David R. Gifford and returned with over 1,000 barrels of sperm whale oil.

Kennebunk-built *Constitution* went twice to the south Pacific, starting in 1833, first under Anthony Gifford and then Eseck Gifford. It returned with 3,700 barrels of oil. *Constitution* was stranded in 1840. Bucksport's schooner *Franklin* made two voyages to the Atlantic out of Salem. Its April 1836 voyage under Alexander Newcomb returned with forty barrels of sperm whale oil. It was withdrawn in 1837 after disappointing returns. Portland-built *Margaret* made four voyages to East Cape and the Indian and Pacific Oceans starting in 1836 under Adley Wilcox, twice with Theodore Lyman Wimpenney and Nathaniel Fales Jr. They combined for 6,800 barrels of oil, 2,700 of them sperm whale, and five tons of baleen. *Margaret* was lost off the Society Islands on February 27, 1850. Portland-built bark *Octavia* made ten voyages to the south Atlantic, Pacific and Indian Oceans starting in 1830 under eight different masters. It brought back a combined 10,174 barrels of oil, 4,164 of them sperm whale, and 14,500 pounds of baleen. *Octavia* was sold in Hobart, Tasmania. Kittery's bark *Iris* made six voyages to the Indian,

Pacific and south Atlantic Oceans, as well as the Falkland Islands, starting in 1836. Its July 1844 voyage included twenty-one-year-old Charles Gamon of Naples, Maine. Under William C. Haynes, the crew brought back 1,300 barrels of oil and five tons of baleen. Lubec native William H. Davis shipped in 1858 under Stephen Bolles and returned in May 1859 with 1,223 barrels of oil, 558 of them sperm whale, as well as 6,576 pounds of baleen. *Iris* was sold to Boston owners.[92]

Maine-built 1824 vessels that became whalers include Kittery's *Ann Parry*, which made six voyages to the south Atlantic, Indian and Pacific Oceans, starting in 1833 under Reuben Ray Jr., Charles Swain, James Youngs, Abiel P. Perry and twice with James Dennet Jr. *Ann Parry* brought back 8,852 barrels of oil, 5,572 of which were sperm whale. It was sold to Salem owners. Wells-built *Gold Hunter* made eight voyages to the south Atlantic, Indian and Pacific Oceans, as well as New Zealand, starting in 1832. Its May 1841 voyage out of Fall River under Albert Wood included twenty-one-year-old Mainer Orin Quint of Sanford and returned in July 1843 with 1,500 barrels of whale oil and over five tons of baleen. *Gold Hunter*'s career numbers total 21,900 pounds of baleen, 12,100 barrels of oil, 1,250 of them sperm whale. In 1849, it was withdrawn for California trade. Bath's *Olive Branch* made one four-year voyage to America's northwest coast under Gilbert I. Place and

Whale ship *Canova*. *Penobscot Marine Museum.*

returned in May 1849 with 4,194 barrels of oil, 224 of them sperm whale, and 21,200 pounds of baleen. *Olive Branch* was sold to California trade.[93]

Wiscasset-built 303-ton *Henry Kneeland* made seven voyages to Pacific, Japanese and Arctic waters starting in 1844. Its three-year 1851 voyage included Enoch Carter Cloud, who kept a journal and noted that he shipped because it was a strong, substantial ship with a good reputation, although he noted *Henry Kneeland* was forty-seven years old. The ship sailed mid-August 1851 under William Hathaway Vinal, and by November 19, it doubled Cape of Good Hope amid tremendous seas. Once in the Indian Ocean, the crew faced "terrific squalls of rain now, wind & hail & violent gales accompanied with Thunder & Lightening rage from one end to the other!"[94] By December 9, the ship was off St. Paul Island, and by January 1852, it was in New Zealand before it went to Japanese waters and then the Sea of Okhotsk amid major storms. It was noted that "the storm increased last night to almost a hurricane; the sea was running absolutely 'mountains high' & the trembling, groaning ship as she was tossed as a feather from crest to trough, seemed like some huge monster in the agonies of death!"[95]

Storms stove in the vessel's weather bulwarks, destroyed its bow and waist boats and smashed the starboard boat to atoms, "carrying away the davits & making the most complete wreck I ever saw!" The crew spent October and November in Hawaii and then were back to Japan and the Sea of Okhotsk. Once homeward bound, *Henry Kneeland* made for Cape Horn and arrived the first week of January 1854. Cloud wrote:

> *We were carrying every stitch of canvass which we could spread, when we were struck by a terrific snow squall which threatened to relieve us of our masts! I went up to assist in furling the main-top-gallant-sail, got it furled and found I could not get down! My hands were frozen stiff and I could not take hold of the shrouds. I laid across the to'gallant yard and beat my hands with all my force, in order to start the circulation of the blood. Could not do it! Things began to get desperate! My clothes were freezing, and such was the pain which I was suffering that I really began to feel unconcerned whether I got safely to the deck or fell a mangled corpse! Winding my arms and legs around the shrouds, I commenced a very perilous descent, not knowing or caring whether I got safely down or not! I managed at length to reach the deck and at once, plunged my hands into a tub of cold water. A short time sufficed to bring them to, and then I suffered!! I never felt such pain before and I am quite sure that I am not very anxious to experience the like again!*[96]

The crew navigated around icebergs as the ship rounded Cape Horn and arrived in New Bedford in April 1854 with 3,386 barrels of oil, 222 of them sperm whale, and nearly eighteen tons of baleen. Its 1858 voyage under Benjamin Kelley included Mainers Bernard Edwards and Alfred H. Hardy, both of Rockland. In August 1859, the ship was reported off Icy Cape, but no whales were taken. It returned May 1862 with 2,938 barrels of oil, 125 were sperm whale, and almost fifteen tons of baleen. In July 1864, under John M. Sowle, *Henry Kneeland* was reported lost in the Arctic.[97]

One final 1824 Maine-built vessel that went whaling was Wiscasset's 372-ton *Tamerlane*. Merchantman for twenty-eight years, it then made a dozen voyages out of New Bedford, starting in 1850. Its 1854 voyage under Joshua B. Winslow included Mainers George W. Bachelder of Baldwin and Peter Good of Eastport. In July 1857, off Kodiak Island and having no luck, they sold bomb lances to *Addison*. In the same season, it hit a reef off Kauai but was floated off with no damage other than loss of false keel. The crew returned in June 1858 with 3,671 barrels of oil, 253 of them sperm whale, and thirteen tons of baleen. Mainers David Turnbull of Bangor and Edson Mcalister of Minot shipped *Addison* in 1858 and returned in July 1862 with 2,546 barrels of oil, 405 of which were sperm whale, and over eleven tons of baleen. George Hart of Eastport sailed it in 1862 under Nathaniel P. Gray and returned in May 1865 with 1977 barrels of oil, eighty-three of them sperm whale, and over eight tons of baleen. *Tamerlane* sailed from San Francisco in 1888 as part of the Pacific Fleet known as Shanghaier. It wrecked with loss of most of the crew off Hilo, Hawaii, on February 2, 1892.[98]

Newcastle-built *Adeline* was launched in 1825 and made eleven voyages to the north Pacific and New Zealand for forty years, starting in 1833. Its 1856 voyage under Asa Taber out of New Bedford included Mainer Henry A. Mason of Portland. In 1859, it was reported that *Adeline* arrived in Honolulu with a new son of Taber's. His wife had given birth at sea and the boy was called a young bowhead. The ship returned in June 1860 with 1,403 barrels of oil, 128 of them sperm whale, and almost eight tons of baleen. *Adeline* was sold to owners in Manchester, Massachusetts, and then to foreign buyers. Phippsburg-built bark *Cora* made four voyages to the south Pacific and New Zealand starting in 1833. Its 1841 voyage under Archelaus Baker Jr. included Mainer Robert Watts of Bucksport and returned four years later, in November 1845, with 1,176 barrels of oil, 621 of them sperm whale, and two tons of baleen. *Cora* was withdrawn from service. Hallowell-built *Cassander* made three voyages to the north Pacific, starting in 1841.

In November 1847, it sailed from Providence, Rhode Island, under Henry Winslow and, took on two locals off Africa. On May 1, as it was put back to sea, a fire broke out about four o'clock in the morning. As the crew fought flames, the two locals jumped overboard, where one drowned. As the other was rescued, he confessed that they had set the fire thinking they were going to be taken to America as slaves. *Cassander*'s fire could not be contained, and its crew was put to boats and drifted without food for ten days before being rescued. Two crew members died of starvation.[99]

In 1825, Bath launched 404-ton *Arbella*, built by Johnson Rideout for Clapp & Boynton. Its advertisement read:

> *For Sale. A* SHIP *on the stocks, of the following dimensions: —Length on deck 114 feet, 28 feet beam, 12 feet 3 inches lower hold, 6 feet 9 inches between decks, and will measure about 400 tons, built of the best seasoned materials, copper fastened, and well salted; the timber above light water mark, and knees of white oak and hackmatack; plank above light water, wales, black, paint and shear rail strakes, of white oak; bottom plank of grey oak; southern pine waist, and locust treenails; the decks of three inch hand sawed plank; the upper deck and waist plugged; to have a round house on deck, billet head, and carved stern, and to be finished in the modern New York style. Can be ready to receive the rigging by the first, and finished by the last of November. For further particulars, apply Messrs.* PARKER *&* ATTWOOD*, New York,* ADAMS *&* AMORY*, Boston, or to the subscribers at Bath.*
>
> *CLAPP & BOYNTON*
> *Bath, Sept. 16, 1825*[100]

*Arbella* was soon whaling and made three voyages to the Pacific and south Atlantic, starting in 1830 under George Harris. The crew returned in April 1834 with 2,300 barrels of oil, 200 of them sperm whale. Under Ellis C. Ellbridge, *Arbella* made two other two-year voyages and returned with 3,900 barrels of oil, 380 of them sperm whale. In 1838, it was withdrawn for freighting.[101]

Kennebunk-built *Eagle* made two voyages to the Pacific starting in 1840 and was then condemned at Rio de Janeiro in February 1846. Its December 1840 voyage included Mainers Joseph Reckerd of Hampden and George Watts of Portland. Under Samuel Perry, they returned in April 1843 with 1,909 barrels of oil, 291 of them sperm whale, and nine tons of baleen. Bath-built, 376-ton *McLellan* was a cotton trader and carried passengers from Ireland's Belfast to Baltimore—150 was its limit. It made ten voyages to Desolation Island and Davis Straits between 1846 and 1852. Its 1849 voyage out of New London included Mainer Garmen Cutter of Eden. Under Christopher B. Chapel, the crew returned in October with 600 barrels of oil and six tons of baleen. Mainers Daniel Crist and Charles Kindred were aboard the last voyage, when *McLellan*, under William Quayle, was crushed in ice of Davis Straits on June 20, 1852. Edgecomb-built brig *Pilgrim* made nine voyages to the south Atlantic, south Pacific and Indian Oceans starting in 1839, after which it was sold to California trade. Saco-built brig *Sarah Louisa* made eight voyages to the south Atlantic starting in 1836, until it was abandoned at sea in 1846. Its 1840 voyage started under Ray Green Sanford but ended with Ebenezer Slocum and included twenty-nine-year-old Mainer Joseph McCobb of Boothbay. The ship returned in April 1842 with 361 barrels of oil, 341 of them sperm whale. Bangor's John Low and seventeen-year-old Hugh McClaines of Eastport shipped aboard *Sarah Louisa* in 1843 for its last voyage. The crew accumulated 435 barrels of oil, but *Sarah Louisa* was abandoned at sea in 1846. Bath-built bark *Wolga* made three voyages to the Indian Ocean out of Fairhaven, starting in 1846. Bangor's John Ferris shipped aboard in 1846 under Grafton Luce and returned in June 1852 with 951 barrels of oil, 118 of them sperm whale, and 11,450 pounds of baleen. It was sold to Boston owners in 1859.[102]

Maine-built 1826 vessels that went whaling include Lubec's bark *Amanda*, which made three voyages to the south Atlantic, starting in 1830. Its May 1831 voyage under Ezra Smith included Mainers Jedidiah Merrell of Berwick and green hand Harris Parsons of Phippsburg. The crew returned in February 1832 with 1,100 barrels of oil and five tons of baleen. *Amanda* was condemned at Mahe in 1834. Warren-built *Charles Adams* made five voyages to the south Atlantic and Falkland Islands out of Stonington, starting in 1831. Its first two voyages' combined take was one thousand fur seal pelts and 4,400 barrels of elephant seal oil. Its 1833 voyage under Charles T. Stanton included Portland native Nathaniel Glasier. The ship came back empty. *Charles Adams*'s 1835 voyage to Patagonia and the Falkland Islands, under Captain William Beck, resulted in 2,300 barrels of oil and over eleven tons of baleen. Its final 1836 voyage was under Daniel Carew. *Charles Adams* was lost at the Falklands in 1837, when it was destroyed by fire. East Machias–built bark *George Washington* made thirteen voyages to the south Atlantic and Indian Oceans starting in the 1830s. Its 1840 voyage under Lemuel H. Eldrige included Mainers Nathan Brick of Gardiner and Isaac Libbey of Newfield but returned empty. Its 1845 voyage included

Mainers John Nicholson and Eli J. Winslow. Under Amos Crowell Baker Sr., they returned in April 1848 with 1,043 barrels of oil, all but 2 sperm whale. Four Mainers shipped on its 1851 voyage under master William O. Harps, including John H. Cram of Sanford, Samuel Crogan of Hermon, William Howard of Bath and Hiram K. Pratt of Bangor. They returned in December 1853 with 315 barrels of oil, all but 3 sperm whale. *George Washington* was condemned in 1858. Biddeford bark *Philetus* made seven voyages out of Stonington, Connecticut, to the south Atlantic, Crozet Islands and the Indian Ocean, starting in 1835. Its total haul included 935 barrels of sperm whale oil, 8,805 barrels of oil and almost twenty-nine tons of baleen. *Philetus* was condemned at Mauritius in October 1850. Penobscot-built brig *Vermont* made one voyage to the south Atlantic in April 1841 under Jonathan Martin and returned in January 1842 with 50 barrels of sperm whale oil. It was broken up that year.[103]

One final 1826 vessel to go whaling was eighty-one-foot, Alna-built bark *Brunette*. It made eight voyages starting in 1831 under Henry Cottle, Matthew Poole and Eddy M. Luce Jr. *Brunette*'s career numbers total 2,260 barrels of sperm whale oil. In the fall of 1843, *Brunette* lay idle at Woods Hole, Massachusetts, and was purchased by gunmaker entrepreneur Samuel Colt. He moved the ship to Alexandria, Virginia, where it was temporarily docked. Colt intended to blow it up with his experimental torpedo and hoped to sell his secrets to the government. He therefore needed a great demonstration. Colt renamed the ship *Styx* and invited President John Tyler and other dignitaries to attend on April 13, 1844:[104]

> *As she approached the spot where the buoys had floated, an explosion took place, and the water was thrown up in a pyramid, but a few yards ahead of her.…The ship held on her course, and in a few minutes another mountain of water, larger and blacker than the first, rose on her larboard bow, and so close to her that she rocked under the undulation…when a third explosion took place—the bows and bowsprit of the ship, instantly shattered into atoms, were thrown into the air. The fore part of the vessel was lifted up almost out of the water, and then immediately sank, while the stern continued above water, and the mizzenmast was left still standing, though in an inclined position. The spars and sails hung in confusion, being suddenly blackened by the smoke, and the whole presenting a wreck in the highest degree picturesque.…A momentary pause of gratified suspense took place, and then the shores resounded with heartfelt plaudits, subsiding into long-continuing murmurs of admiration.*[105]

Due to difficulties with the patent office, hesitation from the war and navy departments and congressional reluctance to pay his asking price, Colt's secret system remained secret, and he took it with him to the grave. *Styx* lay a rotting hulk in the Potomac River—not entirely destroyed, even after subsequent demolition efforts. *Styx* continued to obstruct passage and built up significant sandbars that hampered navigation in and out of Navy Yard until the start of the Civil War.[106]

Bath-built *Avis* launched in 1827 and made one voyage in 1841 out of New London to the Indian Ocean under Gilbert Pendleton. Eight hundred barrels are credited to the ship, but it was lost at sea. Wiscasset brig *Dwight* went straight into whaling and made three voyages to the Pacific and Cape of Good Hope before being withdrawn from service. Its final voyage netted 600 barrels of oil, 200 of them sperm whale, and three thousand pounds of baleen. Bath-built bark *Harriet* carried rum and sugar from St. Croix to Boston but then made one four-year voyage under James Durfee from 1844 to 1848. It went to the Indian Ocean and brought back 1,250 barrels of whale oil and was condemned at Pernambuco in August 1848. Brunswick's 295-ton *Brunswick* was built by Jacob Skolfield. It was one hundred feet long, twenty-five feet and eight inches wide and twelve feet and ten inches deep with two decks. With a square stern, no galleries and a billet head, it cost $16,400 and went into transatlantic trade. After being sold to owners in Providence, Rhode Island, they converted it to a whaler and made five whaling voyages starting 1834. It brought back 8,080 barrels of oil. Sold to New Bedford in 1843, it made three more voyages. In 1848, under Thomas W. Johnson, it went to the Amsterdam Islands and netted 2,145 barrels of oil and fifteen tons of baleen. *Brunswick* sold again and did four voyages out of Dartmouth, Massachusetts, starting in 1851 and took in 4,245 barrels of oil and over twenty-three tons of baleen. Its final voyage was out of New Bedford to Arctic waters in 1862, where it collected 1,260 barrels of oil and twenty-five tons of baleen. On June 27, 1865, *Brunswick* collided with a large cake of ice and stove in its hull below waterline. Listing severely, the crew made plans to abandon it, but the arrival of Confederate raider *Shenandoah* put it to the torch.[107]

One final 1827 Maine-built vessel to go whaling was Portland's *Coriolanus*. By 1844, it was sold to buyers in Mystic, Connecticut. Its first voyage departed Mystic in October 1844 under Gustavus A. Appelman. *Coriolanus* sailed to America's northwest coast and returned in July 1847 with 1,070 barrels of oil and 5,000 pounds of baleen. It sailed in September 1847 under John McGinley for the Crozet Islands and returned in July 1849 with 1,700 barrels of oil and 13,000 pounds of baleen. McGinley took the ship again in 1849 and hauled in almost 1,800 barrels of oil and 25,000 pounds of baleen. Two voyages to the north Pacific netted almost 3,500 barrels of oil and nearly twenty tons of baleen. In its final two voyages, *Coriolanus* returned to the Indian Ocean under William Ellery Nash and brought back 1,607 barrels of oil and 1,700 pounds of baleen. In 1860, Captain George S. Fish found only 93 barrels of oil, and *Coriolanus* was condemned at Mauritius in November 1861.[108]

Maine-built 1828 vessels to go whaling include Bath-built bark *Emma*, which made six voyages to the Atlantic, Indian and Pacific Oceans starting in 1839. On October 15, 1853, it was burned off Paita by its own crew. Kennebunk-built *Fenelon* made eleven voyages to the south Atlantic, south Pacific and Indian Oceans, starting in 1832. Its career totals included 1,539 barrels of sperm whale oil, 13,837 barrels of oil and almost eleven tons of baleen. It was condemned and sold at St. Catherines in Brazil in 1848. Bath-built *Marcia* made eleven voyages to all grounds starting in 1832. One source states that it was eventually sold to foreign owners in 1863, but another source claims *Marcia* as part of second Stone Fleet in the Civil War. Thomaston's *Majestic* entered whaling in 1842 and traveled to the north Pacific, Indian and Atlantic Oceans. Its November 1848 voyage, under Worthen Hall, included Mainers Hugh McDonald of Sullivan, Thomas Percival of Vassalboro and William E. Starrett and Wilder McMitchell, both of China. They returned in April 1851 with 3,073 barrels of oil, 55 of them sperm whale, and over twenty-four tons of baleen. In May 1858, it was reported operating in the Bering Sea and was evidently lucky, as a report noted that it was seen boiling down blubber. Its captain, Job Macomber, took sick and was found delirious, although some noted that it was due to alcohol, as he drank "until he knew not when to stop." Macomber was dead within days, and Alfred C. Chester replaced him. Its last whaling voyage in June 1861, under Alexander Augustus Tripp, included Mainer James R. Norton of Buckfield. The crew returned in November with 158 barrels of sperm whale oil. *Majestic* ended its career as part of the second Stone Fleet.[109]

Falmouth's 1828 bark *Pleiades* made four voyages to the south Atlantic, Indian and Pacific Oceans, starting in 1838 and 1848. Its career numbers total 1,503 barrels of sperm whale oil, 5,420 barrels of oil and eight tons of baleen. It was sold to New Bedford owners in 1849. Freeport brig *William and Joseph* made four voyages to the Atlantic starting in 1837. Its totals include 160 barrels of sperm whale oil. On October 21, 1841, it was lost at sea. Kittery built *William Badger* started whaling in 1845. Its career totals include

2,935 barrels of sperm whale oil, 2,598 barrels of oil and 11,250 pounds of baleen. *William Badger* was withdrawn from service, sold to the U.S. government in 1861 as stores-ship and then sold once again at Wilmington, North Carolina, in 1865. Castine-built *Lucas* made eight voyages out of New Bedford to the south Atlantic and Indian Oceans, New Zealand and the Crozet Islands, with career totals of 1,010 barrels of sperm whale, 13,536 barrels of oil and sixteen tons of baleen. *Lucas*'s 1843 voyage to the Crozet Islands was under Captain Philip H. Borden. The crew had 1,700 barrels of oil aboard when the ship was lost at Fort Dauphin near Madagascar on March 9, 1845. The crew was able to save 900 barrels.[110]

Maine-built 1829 ships that went whaling include Falmouth's bark *Cherokee*, which made fifteen voyages, starting in 1836, to the south Atlantic, north Pacific and Indian Oceans, as well as New Zealand. Its career totals include 3,981 barrels of sperm whale, 16,686 barrels of oil and over fifty-eight tons of baleen. *Cherokee* was sold to New York owners in 1872 and then sold to foreign owners in 1873. Falmouth-built *Montpelier* made six voyages to north the Pacific and Indian Oceans, the northwest coast of America and New Zealand starting in 1839. Its career totals included 1,907 barrels of sperm whale, 13,553 barrels of oil and over sixty-four tons of baleen. *Montpelier* was condemned at Honolulu in March 1857. Freeport's *Condor* immediately went into whaling and made sixteen voyages to the south Atlantic, Indian and Pacific Oceans. Its 1829–30 and 1830–31 voyages were under Portland native Edward Merrill. *Condor* spent its last three voyages in brutal Arctic conditions of extreme fog that made the crew tack, retack and re-retack to keep clear of ice. Its career totals include 3,733 barrels of sperm whale oil, 36,265 barrels of oil and almost 105 tons of baleen. *Condor* was sold in Honolulu in December 1858. Newcastle-built, 512-ton schooner *Two Brothers* should not be confused with the earlier *Two Brothers* lost in the Pacific in 1823. This vessel made nine voyages out of Nantucket with career totals of 332 barrels of oil, 303 of which were sperm whale.[111]

4.

# GOLDEN AGE

*The fishermen know that the sea is dangerous and the storm terrible, but they have never found these dangers sufficient reason for remaining ashore.*
*—Vincent Van Gogh*

By 1830, American whaling had entered its golden age, which would last until the Civil War. Bath-built, 309-ton ship *Lucy Ann* was launched, and its figurehead was carved by noted Bath wood carver Freeman H. Morse. With three-masts, this merchantman was bought by Wilmington Whaling Company in Delaware as a community-led whaling enterprise like Wiscasset or Portland. It sailed to the south Atlantic and Indian Oceans and New Zealand. In 1844, sailor John F. Martin kept a journal, which at some point went into the possession of Charles Gunther, wealthy candy maker, history enthusiast and dime museum owner. In 1920, the Chicago History Museum received Gunter's estate of manuscripts and artifacts and rich materials in seventeenth-century, eighteenth-century and Civil War history, including Abraham Lincoln's deathbed. Chicago History Museum representatives deaccessioned Martin's journal decades ago, and it now resides at New Bedford's Whaling Museum, where in 2016, it was republished as *Around the World in Search of Whales: A Journal of the Lucy Ann Voyage, 1841–1844* by John F. Martin, edited by Kenneth R. Martin.

At the end of his eight-year whaling career, Martin succinctly wrote, "Cursed whaling and quit it. Damn them [who] will not get up after night and burn their shirts to make a light to curse a whaleship."[112]

*Lucy Ann* whaling 1841 watercolor by John F. Martin. *New Bedford Whaling Museum.*

*Lucy Ann*'s five voyages for Wilmington netted 7,800 barrels of whale oil, 1,200 barrels of them sperm whale, and twelve tons of baleen. In 1844, *Lucy Ann* was sold to owners in Greenport, New York. It brought back another 6,800 barrels of oil and almost twenty tons of baleen before it was condemned at Rio de Janeiro in 1850.[113]

In 1830, the Brunswick-built *Romulus* was launched and became a whaler out of Mystic in 1842. Its total from six voyages included 12,739 barrels of oil, 510 of them sperm whale, and 55,191 pounds of baleen. *Romulus* was sold in 1860 to New York owners and operated out of Cold Spring Harbor, where it is suspected that it was used as a slaver.[114]

In 1831, Camden brig *Emeline* was launched. It made six voyages starting in September 1837 under Thomas F. Lambert, which included Saco native William Fuller. The crew returned in July 1838 with 115 barrels of oil, all but five sperm whale. *Emeline*'s last voyages sailed from New Bedford. Topsham-built bark *Curtis* made three voyages out of New London, with no records of any catch. Newcastle-built bark *Junius* made three voyages to the Pacific and Indian Oceans, starting in 1842. Its career totals include 1,972 barrels of sperm whale oil. On October 21, 1851, *Junius* was lost in the Mozambique Channel. Bath-built *Jeannette* made six voyages from New Bedford to the north Pacific starting in 1842. Its October 1858 voyage under Hudson Winslow included Timothy Keith of Portland and returned in August 1862 with 1,892 barrels of oil, 130 of them sperm whale, and over eight tons of baleen. In 1862, it was sold to New York owners. Addison-built brig *Lebaron* made six voyages to the south Atlantic and Indian Oceans starting in 1837. Its totals include 2,614 barrels of sperm whale oil. In 1851, it was destroyed by fire in the Indian Ocean. Thomaston-built schooner *Mac* made two voyages out of Salem starting in 1836. One came back empty, and the second returned with only 70 barrels of sperm whale oil. *Mac* became part of the Civil War Stone Fleet. Maine-built bark *Noble* made six voyages to the south Atlantic, the south Pacific and New Zealand starting in 1837. Its career totals include 7,455 barrels of oil, 725 of them sperm whale, and over eleven tons of baleen. *Noble* was condemned at Auckland in 1846 and entered merchant service. Phippsburg-built bark *Pantheon* made four voyages out of Fall River starting in 1835. Its final 1845 voyage under Francis Dimon included Mainers C. Covell of Woolwich and C.J. Robbins of Pittston and returned in April 1849 with 1,400 barrels of oil, fifty of them sperm whale, and thirteen thousand pounds of baleen. In 1849, *Pantheon* was sold to California trade. Falmouth-built, 263-ton bark *Harvest* made nine voyages over a thirty-year career. Its final voyage in 1859 under David R. Gifford included Mainer William N. Bates of Livermore. The crew collected 360 barrels of sperm whale oil but shipped them by other means, as *Harvest* was condemned at Mauritius in 1861.[115]

In 1832, Franklin-built schooner *Armadillo* was launched. Its July 1866 voyage under Charles H. Hagar included twenty-two-year-old Mainer William Jones of Belfast. *Armadillo* was lost off St. Eustacia in 1867. Brunswick-built bark *Clement* made five voyages to the Pacific and Indian Oceans starting in 1841. Its August 1843 voyage under Luther Fuller included Bath native Isiah Williams and returned with 2,000 barrels of whale oil. Four Mainers shipped aboard its 1846 voyage—three Eastport natives, William Dawns, Henry Woodard and Kenneth McKendre, as well as Benjamin Stafford of Portland. Under Orlando Alexander Lane, the crew returned in May 1849 with 2,400 barrels of oil, 400 of them sperm whale, and four tons of baleen. In 1854, *Clement* was sold to Provincetown owners. Newcastle-built *Copia* made seven voyages to the Indian Ocean, New Zealand and the Arctic. Its career totals include 13,367 barrels of oil, 1,547 of them sperm whale, and over forty-four tons of baleen. *Copia* was condemned at New Bedford in 1855. Wells-built brig *LaGrange* made twelve voyages to the Atlantic, Pacific and Indian Oceans starting in 1837. Its 1840 voyage out of Mattapoisett, under master Albert Daggett, included Mainer David Lewis of Boothbay and returned in May 1841 with 600 barrels of sperm whale oil. *LaGrange* was condemned in 1855 and wrecked in 1856. Belfast-built brig *Malta* made six voyages to the Atlantic, Pacific and Indian Oceans starting in 1845. Its career totals include 2,400 barrels of oil, all but 111 of them sperm whale, and six hundred pounds of baleen. *Malta* was condemned and sold at Fayal on August 27, 1857.[116]

More 1832 Maine-built vessels that went whaling include Brunswick's schooner *Edward*, though another source says it was built in Harpswell for merchant Noah Hinckley. Built by Jacob Skolfield, 133-ton *Edward* was just over seventy-three feet long with one deck, a square stern and billet head. It went into West Indies trade and then was altered into a brig in 1834. In 1841, it made four voyages out of Mattapoisett to Atlantic waters. One was under Matthew Mayhew, two with Joseph R. Taber Jr. and one with Thomas Jefferson Southworth. These were considered "plum numbers" voyages for sperm whaling and returned with 1,870 barrels of oil, 1,630 of them sperm whale. Though it was withdrawn from service in 1846, by 1847 it was back in merchant trade. On April 28, 1848, *Edward* collided with a schooner off Cape Henlopen, Delaware, and sank in ten minutes. Its crew was saved. Robbinston-built schooner *Geneva* sailed from Newport, Rhode Island, to the Falkland Islands in 1836 under a master named Paddock and returned with 900 barrels of oil. In 1837, it was withdrawn from service and sold to New Orleans owners. Saco-built brig *Imogene* went straight into whaling and made eight voyages to the south Atlantic and Indian Oceans, as well as the Cape of Good Hope. Its career totals include 3,480 barrels of oil, 3,030 of which were sperm whale. It was withdrawn from service in 1844. Eastport-built brig *Peri* made nine voyages to the Indian and Atlantic Oceans starting in 1840. These trips included Mainers Lemuel H. Murch of Starks and Thomas Power of Hallowell. *Peri*'s career totals include 6,059 barrels of oil, 5,204 of which were sperm whale, and three thousand pounds of baleen. It was sold at Mauritius in 1863. Kennebunk-built bark *Vermont* also went straight whaling and made seven voyages, mostly from Poughkeepsie, New York. Its career totals include 13,700 barrels of oil, 1,900 of them sperm whale, and nineteen tons of baleen. It was lost at St. Paul Island in October 1847, although the captain and crew survived. Freeport-built, 368-ton *Emily Morgan* went whaling in the Pacific and Arctic for forty years. The *New York Times* recorded it as *G. Morgan*. Owned by J. & W.R. Wing & Company, its career totals include 20,629 barrels of oil, 10,336 of them sperm whale, and nearly forty-eight tons of baleen. In 1871, it was abandoned and lost off Point Belcher, Alaska, in the Arctic Disaster of 1871.[117]

Another Maine-built 1832 vessel that went whaling was Bath's *Roman II*. Started as an Atlantic packet ship, it was considered to be finest of its day: "A new and elegant copper-fastened ship of 350 tons, called the *Roman*, built by Mr. Winthrop G. Farrin, was launched from the yard of J.H. McLellan & Co., on Wednesday morning last. She is owned by Messrs. Winter, Crabtree & Weeks of Portland and is intended for a packet between Savannah and Liverpool, to be commanded by Capt. Weeks."[118]

*Roman* entered whaling in 1835 and made nineteen voyages over forty years. Its first two, out of New Bedford, were to the Indian Ocean. Its August 1840 voyage under master Alexander R. Barker included Mainer Rufus Sproul of Augusta and returned in February 1842 with 3,377 barrels of oil, 418 of them sperm whale, and nearly fifteen tons of baleen. Two Mainers, George R. Blackwell of Georgetown and Alexander Lafferton of Eastport, sailed its 1842 voyage, again under Barker, and came back in July 1844 with 3,301 barrels of oil, 276 of them sperm whale, and fourteen tons of baleen. Its 1844 voyage, again under Barker, included Mainer Richard Marble of Harmony and returned with almost 9,500 barrels of oil and nearly thirty tons of baleen. In 1861, it was withdrawn for merchant service and sold to the U.S. government as a storeship but went back to whaling after war. By 1866, it was again in the Desolation Islands. In 1876, *Roman II* was sold to New Bedford owners and made one final whaling voyage in 1877 under Captain William H. Kelley to the north Pacific. It was lost near Magdalena Bay in 1878. Its total career numbers include 40,285 barrels of oil, 3,161 of them sperm whale, and over seventy-seven tons of baleen.[119]

In 1833, Falmouth-built *Hamilton* was launched. It made eight voyages to the south Pacific, south Atlantic, Crozet Islands and America's northwest coast starting in 1835. Its career totals include 19,205 barrels of oil, 1,235 of them sperm whale, and over eighteen tons of baleen. The ship was condemned at Hong Kong in 1849. Eastport-built, seventy-three-ton schooner *Franklin* made more than twenty-five voyages over sixty years starting in 1842. Its 1846 voyage under Lucius Lothrop Butler included Eastport residents Rory McLane and James Murphy and returned in July 1847 with 450 barrels of oil, 50 of them sperm whale. Its 1847 voyage under Samuel Norrie to the Crozet Islands included Mainer John Colman of Eastport and returned in August 1849 with 25 barrels of oil. Its July 1859 voyage, started under Edwin Church before he left and was replaced by a master named Chappell, included Evan Thomas of Eastport and Alexander L. Dexter of Winslow. They returned from Desolation Island with 974 barrels of oil, 474 of them sperm whale. Its June 1863 voyage to Frobisher Strait under Sidney O. Buddington included Mainer Robert Littlefield and returned in September 1864 with 341 barrels of oil and 5,800 pounds of baleen. *Franklin*'s 1869 voyage to Cumberland Inlet started under George Keeney but ended again under Chappell. John Tafficon of Portland and C.H. Tanaan of Bangor were on board. They returned in October 1870 with 473 barrels of oil and over four tons of baleen. Calais native Thomas M. Near shipped *Franklin* in 1873 to the South Shetland Islands under Charles W. Chester and returned in May 1874 with 267 barrels of oil. Northport's George H. Pomeroy shipped aboard *Franklin* in 1874 under James Waterman Buddington and returned in April 1875 with 160 barrels of sperm whale oil. Following many more voyages, *Franklin* finally returned to New Bedford in July 1891 after a long, illustrious career and was condemned and broken up in 1893.[120]

Another Maine-built 1833 vessel that went whaling was Portland's *Tiger*, which made four voyages to America's northwest coast, as well as the north Pacific and Indian Oceans, starting in 1845. The crew encountered local Inuit near Plover Bay in June 1849, when the captain's wife, Mary Brewster, was awoken by one of the officers shouting that the "bloody Indians" were coming. Trade commenced, and the most prized item was tobacco. A day later, eleven Inuit watercraft escorted *Tiger* to anchor.[121] Mary's husband's predilection to not hunt whales on Sabbath proved a problem not only for his crew but also for her. She wrote:

> *What a time it was for remembering the Sabbath. Perhaps* [the crew] *kept it better than I did, for my only trouble and thought was—I was afraid they would not have the whale....I cannot see how a whaling master can, in the midst of whale*[s], *have sufficient firmness to resist the temptations of trying to get them. I cordially believe there is not a man in the business (how much he may feel disposed to keep the Sabbath) if amongst whale*[s] *wishes the day would pass so Monday would come, and during the day his mind is more taken up with what he is losing than in hoping to keep the day according to God's requirements.*[122]

Mary Brewster became the first Western woman to pass the Bering Strait. Poking northward on the word of an earlier captain, *Tiger* hoped to find numerous whales. New waters and new dangers brought suspense. Mary wrote, "I pray next week may be a little more greasy so these scowling brows may look smooth." They went after walruses, and she wrote that one had come right up to the ship to look at them, and when her husband put an iron into it, "it looked so innocent he left it and was not sorry." *Tiger*'s career totals include 8,943 barrels of oil, 311 of them sperm whale, and almost fifty-four tons of baleen. *Tiger* was withdrawn and sold in 1858.[123]

More 1833 Maine-built vessels included Hampden schooner *Bolton*, which made seven voyages out of Stonington to the Falklands, the Pacific and the Crozet Islands. Its career totals include 5,050 barrels of oil, 1,700 of them sperm whale, and almost two tons of baleen. *Bolton* was sold to Boston owners in 1849. Waterville-built brig *Somerset* made two voyages for New London to the south Pacific starting in 1839. Its first voyage under Benjamin F. Ash included Mainer William Bailey and returned in 1840 with fifty barrels of whale oil. Its next, under William Beck, included Maine natives John Kelley and William Nugent, as well as William Williams from Bangor. It came back empty and was sold at Rio de Janeiro. South Berwick–built brig *Leonidas* made seventeen voyages to the Atlantic and Pacific Oceans. Its 1841 voyage out of Fall River under Ensign Baker included Mainer James Rogers of Eastport and returned in May 1842 with 350 barrels of sperm whale oil. Four Mainers sailed its 1842 voyage under Baker. They were Zanas Arey of Vinalhaven, boat-steerer Nelson F. Paul of Lewiston and William Shitton and Robert Eden, both of Portland. They returned with 265 barrels of oil, all but fifteen sperm whale. Mainers

John Johnson of Leeds and Jesse M. Watson of Fayette were aboard in 1843 under John C. Marble and returned in June 1845 with 260 barrels of sperm whale oil. Its 1845 voyage, under Peleg Cornell, included Mainers Chancy Bonney of Turner and Francis Brown of Brunswick and returned with 230 barrels of oil, 200 of them sperm whale. Portland's Joseph S. Chamberlain was aboard the ship in 1855 under Asa Grinnell and came back with 409 barrels of oil, all but 50 sperm whale. *Leonidas*'s 1857 voyage, under first Rescom Borden until he left and it went under the control of P. Durfee, included Mainer James Nolan of Eastport. Bath resident James Ross was the first mate in 1860, under James L. Skiff, and returned with 388 barrels of oil, all but 70 sperm whale. Rerigged as a bark, *Leonidas* sailed in 1865 under Ebenezer Cook, with Frankfort resident George S. Carris, and returned in July 1867 with 701 barrels of oil, 275 of them sperm whale, and half a ton of baleen.[124]

In 1834, Westbrook bark *Mary and Martha* was built. It made five voyages to the south Atlantic, north Pacific and Indian Oceans starting in 1835. Its career totals include 6,458 barrels of oil, 675 of them sperm whale oil, and over ten tons of baleen. It was condemned and broken up at Buenos Aires in 1855. Kennebunk-built bark *Venice* made six voyages out of New London to the northwest coast of America, the north Pacific and the Indian Ocean, starting in 1844. Its 1847 voyage, under Franklin B. Harris, included Maine natives Robert Carr and John Mason, as well as Bangor resident William H. Thompson and returned with 2,650 barrels of oil, fifty of which were sperm whale, and over seventeen tons of baleen. *Venice* was withdrawn from whaling in 1859 and sold in 1863 to owners in Calcutta, India.[125]

In 1835, 254-ton Nobleboro bark *Cossack* was built and made eight voyages to the northwest coast of America, the Arctic and the Indian and Pacific Oceans. *Cossack*'s mate once recorded that when it drifted into ice, "it made her groan." Maine brothers George and Joseph Green of Thomaston were part of the 1840 crew, as was Elisha W. Parker of Farmington and Daniel L. Randall of Vassalboro. Under master Isaac Delano out of Sippican, Massachusetts, they returned in May 1843 with 1,706 barrels of oil, 350 of them sperm whale, and nearly six tons of baleen. Its 1857 voyage, under John C. Haskins, included Mainer Harvey S. Simmons of Strong and returned March 1861 with 556 barrels of oil, 79 of them sperm whale, and over two tons of baleen. After that voyage, it became part of the Stone Fleet. Scarborough-built, 270-ton *North America* went straight into whaling and made three voyages to the south Atlantic and Indian Oceans out of Wilmington, Delaware. It was lost at Geographic

Bay, Washington, in 1890. A different whaler named *North America* included Mainers Joseph Crisp of Yarmouth, Walter Sackett of Russell and James Thompson of Wiscasset as part of its 1842 crew. Its 1844 voyage to the Pacific included Mainer John Trussel, Bath native Joseph Ella and Augustus Mimsine of Portland.[126]

In 1836, 303-ton Portland ship *Nahant* went straight into whaling, made one voyage and returned with 2,330 barrels of oil, 230 of them sperm whale. Thomaston-built brig *Zoroaster* made six voyages to the Atlantic and Gulf of California starting in 1839. Its 1843 voyage, under Joseph Seabury, included two Mainer seamen, Nathaniel Robbins of Fairfield and Moses Tolman of Industry. They returned with 246 barrels of sperm whale oil. Kennebunk-built *Charleston* made one voyage to America's northwest coast in 1844 and returned with 3,200 barrels of oil, 50 of them sperm whale, and fifteen tons of baleen. *Charleston* was sold in 1847. Nobleboro-built schooner, 111-ton *Tyleston* made thirteen voyages to the Pacific and Atlantic Oceans starting in 1839. Its career totals include 756 barrels of oil, 616 of them sperm whale. Some voyages came back empty or with less than 25 barrels, and one voyage brought back only 6. In 1854, it got to Cape Horn but encountered gales of wind and returned to Pernambuco where it was condemned. Bath-built bark *Cervantes* made two voyages from New London to the south Atlantic and south Pacific starting in 1841. Its first, under Benjamin F. Brown, included Mainer Joseph Clark of South Berwick. *Cervantes* returned in May 1843 with 1,000 barrels of oil and five thousand pounds of baleen. Its 1843 voyage, under Sylvanus H. Gibson, included Eastport native Salter Blowers and sailed in company with *White Oak* and *Stonington*, visiting the Crozet Islands. On *Cervantes*'s return journey from its 1843 voyage, it wrecked off New Holland near Australia on June 29, 1844.[127]

In 1838, Frankfort brig *Montezuma* was built. It made fourteen voyages to the south Atlantic, Indian and Pacific Oceans starting in 1840. Mainer Joseph E. Mercer of Hancock sailed its 1856 voyage under Dennis D. Baxter, and returned with 497 barrels of oil, 398 of them sperm whale. William Roach of Bangor was aboard for its 1858 voyage under master Shubael S. Spooner. *Montezuma* was lost in the Gulf Stream in 1859. Kennebunk-built *General Scott* made five voyages to the Indian Ocean and the north Pacific starting in the mid-1840s. Its July 1848 voyage, under George C. Harris, included Eastport natives William Holland and John Peterson. Charles Brown of Portland shipped aboard it in 1851 under Henry T. Smith. *General Scott*'s 1855 voyage included Monmouth natives and brothers George R. and Epaphras K. Parsons. Its career totals include 9,877 barrels of oil, almost 9,000 of them sperm whale, and over thirty-eight tons of baleen. It was sold to Boston owners in 1862. Kennebunk-built, 420-ton bark *Laurens* went into whaling in 1841 under master Atkins Eldridge. It likely engaged in slave trading, like many other whalers at this time. *Laurens* was seized off the coast of Brazil near Rio de Janeiro by Commodore George W. Storer, commander of U.S. naval forces. Eldridge was charged with fitting out for the slave trade. *Laurens* made two other voyages to America's northwest coast out of Sag Harbor, New York, in 1845 and 1849. Its career totals include 1,400 barrels of oil. Waldoboro-built bark *Elizabeth* made three voyages to the Indian and Pacific Oceans starting in 1841. Mainer Ivory D. Hutchins shipped aboard that voyage, first under Bradford W. Winslow, who was killed at some point. The crew returned in March 1844 with 1,000 barrels of oil, 150 of them sperm whale, and 8,500 pounds of baleen. Its 1844 voyage under Elihu Gifford included Mainers Thomas Percival of Vassalboro, James Spear of Bangor and Oliver Wilson of Shapleigh. *Elizabeth* burned off Fiji in 1846.[128]

In 1839, Portland bark *Sarah* was built and made three disappointing voyages starting in 1845. Its career totals include only 1,300 barrels

Painting of *Montezuma* underway with full sails. *Maine Maritime Museum.*

of sperm whale oil, which it collected during its first voyage. It was condemned at Callao in 1852. Hampden-built bark *Millinoket* made four voyages to the Atlantic, Pacific and Indian Oceans between 1848 and 1862. Its August 1855 voyage, under master Charles A.M. Taber, included Eastport residents John Naughton and D.W. Sumer. The crew returned in September 1858 with 511 barrels of oil, all but two of them sperm whale. Orrington-built schooner *Tamoree* made at least one whaling voyage out of Montville, Connecticut, in April 1856 under master Joseph Comstock, but there are no records of its catch.[129]

In 1840, Orland's 112-ton schooner *Glendower* was built. It was bought from Surrey owners in 1862 and made one whaling voyage to the Atlantic in June 1863 under Nehemiah West and was condemned overseas the following year. Freeport-built bark *Globe* also made only one whaling voyage—this one to America's northwest coast from 1845 to 1849 under master William A. West. Mainer John Dunham of Fryeburg was part of its crew. They caught 3,100 barrels of oil, 100 of them sperm whale, and almost three tons of baleen, before *Globe* was condemned at Valparaiso in 1849 and refitted for California trade. Topsham-built bark *Highlander* also made only one whaling voyage, this one to the Pacific in 1845 under Reuben Cleveland. It returned in 1849 with 600 barrels of sperm whale oil and was condemned at Talcahauno that same year.[130]

In 1841, Richmond bark *Alice Frazier* was built and made three voyages to the Arctic starting in 1851. Mainer John Howard of Knox shipped aboard it in September 1859 under Washington T. Walker. *Alice Frazier* was lost in the Okhotsk Sea in January 1860. Waldoboro-built bark *Avola* made three trips to the Indian Ocean, all under Zenas E. Bourne, starting in 1867. That first voyage included Bangor's John B. Loud. In 1869, *Avola* was in Duke of York Islands working its way up the east coast of New Ireland to Tabar. The crew returned with 832 barrels of oil, all but five sperm whale. Its 1870 voyage included Mainer Joseph O. Hadard of Dover. That voyage collected 1,331 barrels of oil, all but fifteen sperm whale. *Avola* was condemned at Mahe on March 5, 1877. Machias-built bark *Margaretta* made three voyages out of Salem starting in 1851. Its career totals include 567 barrels of oil, 500 of them sperm whale. It was sold to California owners in 1855. Pittston-built bark *Sea Flower* made four voyages to the Atlantic starting in 1851. Its final 1856 voyage, under master Sylvanus Cleaveland, included Mainers George L. Heald of Madison and first mate William Ralph of Wells. The crew had collected 213 barrels of oil, all but 11 sperm whale, when *Sea Flower* wrecked off the Cape Verde Islands in 1859. One final 1841 Maine-built ship that went whaling was Bucksport's schooner *Warwick*. Under a master named Grozier, it sailed from Bucksport to western islands, also known as the Outer Hebrides, and returned with 110 barrels of whale oil, before it was withdrawn from service in 1842.[131]

Waldoboro-built bark *George Henry* made four voyages to Davis Straits and Hudson Bay starting in 1855. Its 1860 voyage, under Sidney O. Buddington, included William Ellard of Calais. That season, *George Henry* took Arctic explorer Charles Francis Hall northward and got as far as Baffin Island, where *George Henry* was forced to spend the winter. Hall was informed by local Inuit about relics of Martin Frobisher's sixteenth-century mining venture at Frobisher Bay on Baffin Island. Hall traveled there to collect them. He also found what he believed to be evidence that some of Sir John Franklin's lost expedition might be alive. *George Henry* returned in September 1862 with 564 barrels of oil and over five tons of baleen. It was later wrecked in Hudson Strait on July 16, 1863.[132]

In 1842, Portland-built *American* was launched. It made one voyage as a whaler in 1857 out of New London under master R.B. Baldwin. In 1844, *Edgar* was built in Brunswick. It made one whaling voyage to the Pacific out of Cold Spring Harbor, New York, in 1852 under master Samuel B. Pierson and is credited with 3,459 barrels of whale oil and thirty-three thousand pounds of whale bone but was reported lost in the Okhotsk Sea in 1855. Yarmouth-built *Helen Augusta* made two voyages to the north and south Pacific starting in 1850, until it was burned by its crew on February 15, 1856, in Mangonui, New Zealand. Bath-built bark *Isabella* made four whaling voyages to the Pacific and Arctic starting in 1852. Its August 1859 voyage, under Moses G. Tucker, included Mainer Matthew Kelley and returned in May 1863 with 2,548 barrels of oil, only 26 of them sperm whale, and fourteen tons of baleen. Its September 1863 voyage, under master Hudson Winslow, included Mainer Thomas Hackett of Biddeford. *Isabella* was captured and burned by Confederate raider *Shenandoah* on June 26, 1865. Falmouth-built bark *J.E. Donnell* made four voyages to the Arctic starting in 1845, until sold to Fairhaven owners and broken up in 1857. One final 1844-built vessel that went into whaling was Blue Hill's *Tahmaroo*, which made three voyages to the north Pacific and Atlantic waters starting in 1852. Its career totals include 3,814 barrels of oil, 432 of them sperm whale, and over fourteen tons of baleen. *Tahmaroo* was sold at Fayal in December 1861.[133]

In 1845, brig *Germ* was built in Eastport and made one voyage out of Truro, Massachusetts, under Alexander B. Ryan in 1852. There was no record of any catch, and *Germ* was condemned at St. Thomas on November 4 of that year. Portland-built *Warren* made four whaling voyages to the north Pacific and Indian Oceans starting in 1851. Its 1855 voyage included Mainer Henry Charlton of Calais as *Warren*'s third mate. That voyage went through three captains, first master Preserved S. Wilcox left ship at some point and was replaced by George Huntley, who also left ship and was finally replaced by Albert Miller. The crew returned in May 1860 with 1,596 barrels of oil and nearly seven tons of baleen. *Warren* was withdrawn from whaling service that year and condemned at St. Catherine's in Brazil in March 1861. Portland-built 154-ton bark *Janet* made ten voyages out of Westport, Massachusetts, to the Pacific, Indian and Atlantic Oceans starting in 1846. In June 1849, *Janet* dropped its whaleboats after spotting a sperm whale. One boat, which included *Janet*'s captain, was upset during the chase. By the time it righted, *Janet* and other whaleboats had vanished over the horizon. They drifted for three weeks without any provisions. As men started to die, they sustained themselves by cannibalism. Many were drawn to see who would next be sacrificed for the nourishment of others. By the time they were rescued near Coos Island by another whaleship, only Captain N.A. Gifford and one other crewman were still alive. *Janet*'s 1859 voyage, under master George G. Coffin, included Mainer James Lawlor of Conway. They returned in May 1863 with 835 barrels of sperm whale oil. Portland residents A.H. Johnson and Charles H. Goodrich joined Edward K. Briggs of Freedom aboard *Janet* in 1866 under Alonzo J. Marvin and returned in June 1869 with 735 barrels of oil, 595 of them sperm whale, and seven hundred pounds of baleen. In 1874, *Janet* was sold to New Bedford owners and made three more voyages to the Atlantic. It was abandoned at sea on November 23, 1879.[134]

In 1846, bark *Afton* was built in Eastport. Within five years, it was in whaling and made four voyages to the Atlantic, Pacific and Indian Oceans. Its May 1856 voyage, under master James M. Clark, included Sumner H. Boynton of Bangor. The crew returned in August 1858 with 1,105 barrels of sperm whale oil. John Bryant, also of Bangor, joined *Afton* in 1858 under master Francis Allen. They returned in August 1862 with 740 barrels of oil, all but 2 sperm whale. It was sold to New York owners in 1862. Nobleboro-built bark *Antelope* made five voyages to the Pacific, Davis Straits and Hudson Bay starting in 1851. Its October 1861 voyage, under George Taber, included Mainer Thomas Davis of Carroll and returned with 1,500 barrels of oil and twelve tons of baleen. Its career totals include 3,884 barrels of oil, 681 of which were sperm whale, and nearly nineteen tons of baleen. *Antelope* was lost off Cumberland Inlet in 1866. Another 1846 Nobleboro bark was *James Andrews*, which made three voyages to the north Pacific, Indian and Atlantic Oceans starting in 1851. Its career totals include 3,537 barrels of oil, 782 of which were sperm whale, and four tons of baleen. Searsport-built bark *Osceola III* made eight voyages to the Atlantic and Pacific Oceans starting in 1852. Its 1856 voyage, under master John P. Carr, included twenty-seven-year-old Mainer William Durgin of Saco and returned in September 1858 with 626 barrels of oil, all but fifty-two sperm whale. In 1866, under master Martin Malloy, the crew lowered both waist boat and captain's boat for a lone bull whale. Both struck with irons, but the whale stove both boats with a sweep of its tail and cut the bottom out of the captain's boat. A bomb was fired into the whale without any effect men could see. *Osceola III* came over and picked up the swamped men, and Malloy went after the whale, ready to fire another bomb. As they approached, the whale turned and charged *Osceola III* with its mouth open. The collision made the ship tremble and knocked off its bow's cutwater. Meanwhile, two more lances and a bomb were thrown into the whale. *Osceola III* tried to stay close, but the whale kept off even with a tangle of two tow lines and parts of the stove waist boat. After a twelve-hour battle and thirty-one bombs fired into it, Malloy was finally able to kill the whale, which reportedly yielded 115 barrels of oil.[135]

Portland bark *Henry Trowbridge* was built in 1847 and made two voyages to the Atlantic starting in 1877. Its career numbers include 800 barrels of oil, 600 of them sperm whale. It was abandoned at Fayal on November 12, 1882. Pittston-built bark *Keoka* made two voyages to the Pacific and Atlantic Oceans from 1853 to 1860 and brought back a combined 1,827 barrels of oil, 1,171 of them sperm whale. It was sold for California trade in 1861. Pembroke-built bark *N.D. Chase* made six voyages to the south Atlantic, Indian and Pacific Oceans starting in 1851. Its December 1854 voyage, under a master named Hussey, included Michael Sharon of Portland and three Mainers from Sidney, Henry A. Case, George C. Folger and Aurelius C. Frost. They returned in December 1856 with only forty barrels of oil, twenty-five of which were sperm whale. *N.D. Chase* was sold to Nova Scotia owners in 1861. Augusta-built bark *Winthrop* made three voyages to the Atlantic, Pacific and Indian Oceans starting in 1852. Its July 1859 voyage, under master William Weeks, included James Lewis of Portland. The crew brought back 368 barrels of oil, all but 48 from sperm whale. It was sold to Boston owners for merchant service in 1863 and then sold to foreign operators. Franklin-built *Cleora* made four voyages to the Pacific and Indian

Oceans starting in 1852. Its 1855 voyage, under master Shubael Hawes Norton, included Mainers Frank Dana of Parkman, Cyrus H. Day of York, John F. Jones of Lebanon and Henry Williams of Belfast. *Cleora* brought back 1,513 barrels of sperm whale oil. Castine's Albert Grayham shipped aboard its 1859 voyage, first under master William White then under Isachaar H. Akin. The crew reported no whales caught, and *Cleora* was condemned at Mauritius in July 1862.[136]

In 1848, Damariscotta bark *Delaware* was built. It made three voyages to the north Pacific starting in 1852. That first voyage included Portland native Daniel O. Leary, but *Delaware* went through four captains. Caleb Stong Holt left ship at some point and was replaced by a man named Holman, who also left ship. Theodore Brown replaced him but was then replaced by Isaac Allen, who finished the voyage. The crew returned with 2,866 barrels of oil and over thirty-one tons of baleen. *Delaware* was reported lost off Ballenas Bar in lower California in 1860. Woolwich-built, 140-ton brig *Homer* made five voyages to the south Atlantic starting in 1852. Its career totals include 873 barrels of oil, 449 of which were sperm whale. It was lost off Terceira in the Azores on September 7, 1860. Damariscotta-built bark *Ionia* made five voyages to the Indian and Pacific Oceans starting in 1851. Its 1858 voyage included Mainer Allen Smith of Standish and three masters. Captain James W. Morse died and then was replaced by James M. Russell, who left ship at some point and was replaced by an N. Bonney. *Ionia* was sold to Honolulu owners in 1861. Wells-built bark *Joshua Bragdon* made three voyages to the Pacific and Indian Oceans starting in 1853. Its career totals include 3,346 barrels of oil, all but 617 sperm whale. It was sold to New York owners for merchant service in 1864. Pittston-built *Mary Wilder* made seven voyages to the Pacific starting in 1849. Its 1860 voyage, under master Sylvanus Cleaveland, included Mainers John Breen of Freeport and Samuel S. Smith of Portland. They returned with 430 barrels of oil, all but 50 sperm whale. *Mary Wilder* was sold to New York owners in 1864. Hallowell-built, 224-ton bark *Oregon* made three voyages to the south Atlantic starting in 1853. Its totals include 2,051 barrels of oil, 1,084 of which were sperm whale, and 1,550 pounds of baleen. *Oregon* was sold to Fairhaven owners for west India trade in 1859.[137]

In 1849, 423-ton *Mary Merrill* was built in Robbinston and became a New Bedford whaler, although there is no record of it in NMDL's database. Waldoboro-built bark *Aerial* made at least three voyages out of Fall River to the Atlantic and Indian Oceans, starting in 1851. Its 1853 voyage, under master Philip H. Borden, included Mainers Francis Finlay of Lewiston Falls and William F. Smith of Norway. They returned with 356 barrels of oil, all but 30 sperm whale. *Aerial* was lost at sea in 1859. Robbinston-built brig *Amaret* made six voyages to the Arctic, including the Davis Straits, starting in 1853. Its 1859 voyage to Cumberland Straits, under master John T. Quayle, included Bucksport natives John Forsyth and Edward Smith, as well as George A. Williams of Wells. They returned January 1860 with 150 barrels of oil. *Amaret* was lost in Cumberland Straits later that year. Thomaston-built bark *S.H. Waterman* made one voyage out of Stonington, Connecticut, to the north Pacific in 1851. Under Palmer Hall, it returned in April 1855 with 2,640 barrels of oil and over fourteen tons of baleen before it was withdrawn from service that year. Camden-built bark *Sacramento* made three voyages out of Westport, Massachusetts, to the Indian and Pacific Oceans, starting in 1852. Its career totals include 2,144 barrels of oil, all but sixty-nine sperm whale. Cumberland-built bark *San Francisco* made four voyages to the Atlantic and Indian Oceans starting in 1852. Its 1855 voyage, under master Elisha G. Cudworth, included Mainer John Gleason of Calais and returned with 999 barrels of oil, 276 of which were sperm whale. Charles Anderson of Waterville and Wilder F. Semen of Whitefield joined the crew in 1857 under first Philip H. Omey, who left ship and was replaced by Edwin A. Perry. They returned with 675 barrels of oil, 445 of which were sperm whale. William George of Standish shipped aboard its final voyage in August 1859 under master Daniel F. Worth. The ship foundered off Montauk Point, New York, on February 24, 1862.[138]

More 1849-built Maine vessels include 183-ton Portland bark *Sarah B. Hale*, which made three voyages to the Atlantic. Its 1875 voyage, under master Holder C. Slocum, included Mainer William Stunke of Poland and returned with 680 barrels of sperm whale oil. *Sarah B. Hale* was condemned at St. Helena in April 1880. Robbinston-built bark *Manuel Ortiz* made three voyages to the north Pacific starting in 1851. Its final 1857 voyage, under master James S. Hazard, included Mainers Arthur Patterson of Hollis, Joseph A. Stoddard of Dover and James M. Thistle of Eastport. They returned in June 1860 with 1,973 barrels of oil, 110 of them sperm whale, and almost 13 tons of baleen. *Manuel Ortiz* was sold to New York owners and then foreign owners in 1861. Thomaston-built, 96-ton schooner *Walter Irving* does not appear in the NMDL database and may have made seventeen voyages out of Provincetown to Atlantic over twenty years, until it was sold to New York owners in 1871. In 1850, Robbinston bark 293-ton *Hannah Brewer* was built, but by 1852, it was sold to New

London owners for whaling. It made two voyages to Desolation Island for a combined total of 2,675 barrels of oil, 397 of which were sperm whale, and 7,400 pounds of baleen. It was condemned at St. Helena on February 19, 1857. Jonesboro-built bark *Lady Suffolk* made five voyages to the Indian and Atlantic Oceans starting in 1853. Its career totals include 1,414 barrels of oil, 1,244 of which were sperm whale. It was withdrawn from service in 1860.[139]

Waldoboro bark *Civilian* was built in 1851 and made two voyages out of Provincetown to the Atlantic starting in 1860, both under master Elisha S. Burch. It returned with a combined total of 1,583 barrels of oil, all but 60 from sperm whales. *Civilian* was withdrawn from service in 1864. Calais-built bark *D.M. Hall* went into whaling and made three voyages to the Atlantic and Indian Oceans starting in 1852. Its final October 1853 voyage out of Fall River, under master Spencer Pratt, included twenty-year-old Mainer William Wallace of Portland. The crew returned empty in 1855, and *D.M. Hall* was sold to Papeete owners that year. Wells-built schooner *Florence* made four voyages to the Atlantic and Hudson Bay starting in 1872. Its August 1877 voyage, under master George E. Tyson, went to Cumberland Straits with Portland native Richard B. York but returned empty. Bath-built brig, 347-ton *Alfred Gibbs* went straight into whaling and made five voyages out of New Bedford to the Pacific and Arctic waters. Its June 1865 voyage, under master Edmund E. Jennings, included Mainer Henry Robertson of Portland. The crew returned in September 1869 with 1,529 barrels of oil, all but 180 of them sperm whale, and 972 pounds of baleen. *Alfred Gibbs* sold to New York owners in 1873. Bath-built bark, 215-ton *Helen Snow* also went straight into whaling. Its 1862 voyage started under master Joseph S. Adams Jr., but he died and was replaced by Thomas G. Campbell, who left ship at some point and was replaced by George W. Luce. That voyage out of New Bedford included Mainers William Potter of Fairfield and Charles Smith of Troy and returned with 2,129 barrels of oil, 810 of them sperm whale, and nearly thirteen tons of baleen. *Helen Snow* was abandoned in the Arctic Disaster of 1871. It was salvaged and then sold to Russian owners and renamed *Tugar*. It was later renamed *Desmond*.[140]

More 1851-built vessels include one-hundred-ton Rockland schooner *Flying Cloud*, which sailed as part of Stonington's whaling fleet and operated in Patagonia and south Atlantic waters. Its two voyages resulted in only 40 barrels of oil. In 1853, it accompanied *Sarah E. Spear* and *United States* to the South Shetland Islands. Chelsea-built schooner *Lively* went whaling out of Fairhaven under Vassalboro native captain Job Pierce. It was reported lost with all hands in 1852. Job's brother Eben Pierce, another whaling captain, added Job's name and reference to the loss of *Lively* onto his own gravestone located at Maine Avenue Cemetery in Farmingdale, Maine. Gardiner-built *Hunter* also went straight into whaling and made more than twenty-five voyages over the next forty-two years to the Indian, Atlantic and north Pacific Oceans, as well as Japanese waters and the Sea of Okhotsk. *Hunter*'s June 1859 voyage, under master Alden Besse, included Mainers James Ellsworth of Portland, Edwin Irving of Vassalboro and James Keating of Fayette. The crew returned in August 1863 with 2,677 barrels of oil, 2,042 of them sperm whale, as well as 2,700 pounds of baleen. Mainer James Ellsworth shipped aboard *Hunter*'s 1863 voyage under master Asa S. Tobey and returned with 2,658 barrels of oil, 366 of which were sperm whale, and over nine tons of baleen. William Longley of Portland sailed Hunter in 1871, under master Charles L. Holt, and returned in July 1875 with 3,800 barrels of oil, 2,700 of which were sperm whale. From 1886 to 1890, *Hunter* made voyages to Arctic waters from San Francisco as part of the Pacific whaling fleet known as Shanghaiers. Its prolific career totals include 29,507 barrels of oil, 9,950 of which were

Oil painting of *Flying Cloud*. *Penobscot Marine Museum.*

sperm whale, and 244,650 pounds of baleen. *Hunter* worked from San Francisco until in 1900, when it was lost off Cape Romanoff, Alaska.[141]

Falmouth-built 1851 bark *Robert Morrison* also went straight into whaling and made seven voyages to the north Pacific, Indian and Atlantic Oceans starting in 1851. Its August 1857 voyage, under master Benjamin W. Tilton, included Mainers John Julian of Danville, Robert Marshall of Portland, Alpheus Peaslee of Gardiner and Charles Stevens of Otisfield. The crew returned in April 1861 with 2,051 barrels of oil, 215 of them sperm whale, and nearly thirteen tons of baleen. James Doyle of Portland shipped aboard in June 1861 under master Crary B. Waite and returned in August 1864 with 1,139 barrels of sperm whale oil. *Robert Morrison* was condemned at Sydney, Australia, in December 1864. Robbinston-built schooner *Sea Breeze* made one voyage out of New London in 1855 under master Silas Buchanan but left no record of any catch and may have wrecked off the Cape Verde Islands in 1859. Georgetown-built 1851 bark, 425-ton *Sophia Thornton* went straight into New Bedford whaling, although the NMDL database refers to her as Bath-built. It made four voyages to Pacific waters. Whaling frustrations were readily apparent when its master recorded that "lowered and fastened. The first iron drawed. The second broke. Lost the whale. At 4 PM saw one whale, lowered and fastened. Got foul line and boat stove and line parted. Lost the whale, 3 irons, spade, and some line."[142]

Its April 1860 voyage, under master William P. Briggs, included Mainer Thomas Welsh of Calais. The crew returned in September 1864 with 1,342 barrels of oil, all but 18 sperm whale oil. It was later lost to Confederate raider CSS *Shenandoah*.[143]

There was also the 1851 Gardiner-built, four-hundred-ton bark *Trinity*, which entered whaling in 1868 out of New London under Master Alfred Turner. The ship went to Kerguelen Island in the far south of the Indian Ocean, accompanied by whaler *Roman*. That voyage for *Roman* to Desolation and Heard Islands, under master Jared Jernegan, lasted three years and included Mainers Frank Jones of Biddeford and Bangor native Thomas McNear. *Trinity* remained, whaling around Kerguelen, while *Roman* continued to Heard Island two hundred miles away, where it was crushed in ice on September 7, 1871. *Roman* had long attracted Mainers aboard for whaling voyages. In 1845, Joseph F.G. Moons of Bath shipped aboard, and Andrew J. Estes of Portland sailed it in 1847. James Newlegin of Parsonsfield was part of its 1855 crew, and *Roman*'s 1859 voyage included Mainers John Farnow of Lincoln, Jedidiah L. Farr of Gray, Arnold Gray of Etna and Edward Scarlet of Portland. *Trinity*'s July 1870 voyage included Portland native John Foley and sailed in company with *Flying Fish*, stopping at South Georgia, where *Trinity*'s cooper died and was buried at King Edward Cove on January 10, 1871. Yield was light that season—*Trinity* returned with only 210 barrels of oil. Master Erasmus Darwin Rogers did make geological collections. By 1880, *Trinity*, under John L. Williams, visited Kerguelen Island, where it deposited several casks and spare rigging at Betsey Cove in Accessible Bay on the island's northeast side. This lightened the load while the ship sailed 250 miles to Heard Island. There, it wrecked at Spit Bay on October 17, 1880, and two of its crew froze to death. Thirty-one others were rescued by U.S. Navy corvette *Marion* on January 14, 1882, and they were taken to Cape Town.[144]

In 1852, Calais bark *Pamelia* was built and made two trips to the Indian Ocean, starting in June 1855. The first voyage, under master Edward Coggeshall, included Mainer William Hawes of Bangor and returned in August 1858 with 1,382 barrels of sperm whale oil. It was withdrawn from service and sold to Newport, Rhode Island owners in 1862. Perry-built bark *Rose Pool* made two voyages to the Indian and Pacific Oceans, starting in 1856 under master Alexander P. Fisher. Its combined total includes 2,532 barrels of oil, 1,524 of them sperm whale, and 4,528 pounds of whale bone. *Rose Pool* sold to New York owners in 1864. Cape Elizabeth–built bark *Florence* made six voyages, mostly out of Warren, Rhode Island, and later from Honolulu or San Francisco, mostly to the Indian and northern Pacific Oceans, starting in 1852. Its career totals include 4,268 barrels of oil, 1,318 of which were sperm whale, and 19,075 pounds of baleen. On August 8, 1878, during its last voyage, *Florence* was caught and stove by ice four miles off Point Barrow, Alaska. Robbinston-built, 211-ton clipper bark *Francis* (or *Frances*) *Palmer* went into whaling in 1886 as part of San Francisco's Pacific whaling fleet as a Shanghaier. It made three voyages to north Pacific and Arctic waters. In 1887, under Frederick A. Barker, it conducted trade with Inuit at Plover Bay for five boxes of tobacco, two packages of leaf tobacco, one Winchester rifle, three boxes of reloading tools, two dozen knives, seven hundred WCHF rifle cartridges, three hundred fifty-grain-loaded cartridges, twenty-seven bags of flour, three hundred pounds of bread, twelve packages of matches, four hatchets, two axes, one saw, one dozen thimbles, one dozen spools of thread, one dozen papers of needles and five pounds of beads. That season, it returned with 230 barrels of oil and two tons of baleen. On its last trip, it was discovered stove in with its colors at half-mast. Fellow whaling ships helped fix the ship and it limped back to San Francisco on November 4, 1888.[145]

In 1853, 299-ton bark *Wavelet* was built in Robbinston and put into the California run. It made one whaling voyage in October 1855 under master George Swain and returned in 1860 with 2,714 barrels of oil and over 17 tons of baleen. It was sold in 1860 to San Francisco owners for $16,000. One source says that by 1864, it had Shanghai owners and was still whaling. Robbinston-built 326-ton bark *A. Houghton* went straight into whaling and made three voyages to the Atlantic and Pacific Oceans and Hudson Bay. Its 1853 voyage out of Fall River, under master John C. Marble, included Mainer Charles H. Small of Gray. The crew returned with 1,500 barrels of oil, 700 of which were sperm whale, as well as 1,400 pounds of baleen. Portland's Jeremiah Harrington was part of its July 1857 crew under master Orlando G. Robinson. The ship returned in May 1861 with 825 barrels of sperm whale oil and 648 pounds of baleen. Its 1876 voyage out of New Bedford, under master James G. Sinclair, included Mainers Edward Cronan of Franklin and John F. Marley of Portland. They collected 200 barrels of oil and 4,500 pounds of baleen. *A. Houghton* was sold to California owners and then to the U.S. Navy. Brooklin-built schooner *Graduate* made two voyages to the Atlantic starting in 1868 but was lost at sea the following year. Frankfort-built brig *John Hathaway* made one voyage out of Fairhaven to the Atlantic in 1866 under William H. Haskins but recorded nothing caught. It was condemned at St. Thomas later that year.[146]

More 1853-built vessels that went whaling include Gardiner's 425-ton *Kingfisher*. It went straight into whaling but made only one voyage to the Arctic under Martin Palmer. The crew collected 2,102 barrels of oil, 500 of which were sperm whale, and nearly five tons of baleen. *Kingfisher* was lost off Company's Island on May 13, 1855. Georgetown-built bark *Silver Cloud* made one voyage to the north Pacific starting in 1856 under master Frederick Coggeshall. The crew included Mainer William Riley of Springville. Coggeshall's wife, waiting in Hawaii, died of tongue cancer while her husband and ship were in northern waters. The crew returned with 2,657 barrels of oil, 645 of which were sperm whale, and over eleven tons of whalebone. *Silver Cloud* was sold to New York owners in 1862 for China trade and burned later that year.[147]

In 1855, 175-ton brig *Hidalgo* was built in Machias. It made eighteen voyages to the Pacific starting in 1879. From 1886 to 1890, it was a Shanghaier out of San Francisco. By 1890s, it had new owners but still sailed from San Francisco to north Pacific and Arctic waters. On its last voyage on July 24, 1896, *Hidalgo* was forced ashore by ice off Point Hope. Its career totals include 4,170 barrels of oil, only 40 of which were sperm whale, and over twenty-nine tons of baleen. Robbinston-built, 330-ton bark *Tempest* made one voyage out of New London to the Pacific in 1857, first under master Gurdon L. Allyn, who left ship and was replaced by Asa W. Fish. The crew returned in April 1861 with 2,256 barrels of oil and nearly five tons of baleen. By 1868, it was owned and operated out of Valparaiso. Thomaston-built, four-hundred-ton bark *Architect* made one voyage out of New London to the north Pacific. Its 1857 voyage, under master Asa W. Fish, included Bangor native Joseph Loud and returned with 1,552 barrels of oil and nearly nine tons of baleen. It was withdrawn from service in Honolulu in 1859. Eagle Island–built, seventy-seven-ton schooner *U.D.* made three voyages to the Atlantic and Davis Straits in the Arctic starting in 1866. Its career totals include a disappointing 227 barrels of sperm whale oil. It was condemned and sold at Barbados in January 1870.[148]

In 1856, Rockland's ninety-nine-foot, 193-ton schooner *Pilot's Bride* was built. In January 1880, Captain Joseph J. Fuller purchased it in New York for the new owner, New London whaling magnate C.A. Williams. It was insured for $9,000. Fuller thought it the right size for the treacherous shallow waters of Kerguelen Island. He also liked that it did not need any major repairs: "All she needed was some alterations from a trading vessel to a whaling vessel, such as cutting-in gear and boats and the outfits in general of a whaleman that is fitting for a two years' voyage." It made one whaling/sealing voyage to Desolation Island from 1880 to 1883. With a load of two thousand casks for whale and seal oil and enough salt for four thousand sealskins, *Pilot's Bride* sailed from New London in April 1880, with Fuller's older brother as second mate. By May, it was off the Azores and Cape Verde Islands, where the crew got its first catch, a sperm whale that produced sixty-five barrels of oil.[149] The crew met its first serious weather in the far south Atlantic. Captain Fuller described the intensity of the rain:

> *I thought that I had seen it rain before, but all the rain squalls that I had seen were nothing but scotch mists to this, for it did not come in drops but seemed to come in streams and it lasted about* [a] *half hour.…About eleven o'clock as I was laying down on my bed, I heard* [a] *crash. I sprang from my bed and jumped on deck.…I just had time to see that the port waist boat had gone and as it came aft and got under the port quarter boat it rose on a sea, struck the quarter boat, and carried away boat, davits, and all the tackling that belonged to the boat, making a clean sweep of the port side.…It was blowing heavy gale and a tremendous sea running. When*

> *the schooner would get on top of a sea, she would shoot from the top like shot from a gun…there came along another heavy sea and cut the stern of the boat that we had across the stern; that left us with only one boat.…For about twenty-five hours it blowed* [sic] *a hurricane and the sea looked like mountains coming down upon us but the* Pilot's Bride *made good weather of it by shipping but little water.*[150]

By October, *Pilot's Bride* made landfall at Rhodes Bay on Kerguelen Island. Fuller continued to Norton's Harbor, where he intended to make his headquarters. There, he had crew members break out a small schooner that they carried, land all spare provisions and shooks and cask staves tied in bundles. They repaired their remaining boat and engaged in elephant and leopard seal hunting as they worked around the coast. By August, they were on the windward side and, by the end of October, had taken 800 barrels of oil. By November 20, 1880, they had filled all their available casks with oil. Fuller went back to Norton's Harbor and placed 1,300 barrels there to keep while the ship loaded up its deposited shooks and went in search of seals for the next three months. By the end of January, Fuller was ready to take part of his haul to Cape Town and arrange transport of it back to New London. He loaded 1,100 barrels of oil and 1,700 salted-down sealskins aboard *Pilot's Bride* and, on February 2, 1881, sailed for Cape of Good Hope.[151]

They arrived at Cape Town after a good passage of twenty-nine days but had difficulties with local authorities over permits, permission and a load of potatoes. Fuller finally got away for Kerguelen on April 2 and put in at Betsey Cove. He saw a ship had unloaded piles of its own gear and spare rigging. It belonged to Maine-built bark *Trinity*. Before his departure, Fuller had had discussions with *Trinity*'s owner and learned they, too, planned to operate at Desolation Island. As sealing and whaling were highly competitive ventures, Fuller determined to keep his ideas to himself on where any seal colonies on Kerguelen might be located. Among the items that came ashore from *Trinity* was a note for Fuller from *Trinity*'s master, John L. Williams, that said *Trinity* had arrived the first of August, landed spare material, sailed for Heard Island and asked Fuller to come over and check on them. Since it was now June and *Trinity*'s spare material was untouched, Fuller concluded that *Trinity* had probably come to grief at Heard Island, and indeed it had. He decided to finish at Kerguelen then go to Heard Island and check on *Trinity*, but he never got a chance to render assistance. Off Betsey Cove in Accessible Bay, weather deteriorated to thick snow and strong winds. Fuller, aware of a nearby set of dangerous rocks suitably called Rocks of Despair, tried to get to open sea: "We came about and stood off-shore, heading north northeast by compass and by that time it became so thick and dark that we could not see more than half the length of the schooner ahead."[152] He thought they were getting clear of the reef, but posted lookouts reported seeing kelp on the water and a rock near lee bow. "I jumped to the lee side and caught a glimpse of the rock, which was not off more than half schooner's length, and the schooner was coming up fast into the wind. But it was no use; she could not get clear of the rock. The next minute a sea came and picked her stern up and landed it on top of the rock and as fast as the sea would throw her stern up she would forge ahead and draw off again…third sea carried away the rudder."[153]

*Pilot's Bride* came off that rock and wallowed in twelve fathoms of water. Fuller ordered anchors and chains out. He did not know what was ahead, as visibility was down to zero in blowing snow, and the rock that had shredded his rudder was now somewhere behind them. The ship seemed secure at the moment, so Fuller ordered the crew to take a meal, divided them into three groups and announced they would wait out the night and further assess their situation with morning light. He went on deck a few hours later. In his memoirs, Fuller wrote of that scene: "The wind was still to the northward and blowing strong [with] a heavy sea, the schooner pitching heavy, taking water over the bows, but still she did not seem to be riding heavy to her anchors. I could make out that we was [*sic*] close to the Rocks Despair and that we was laying in the middle of the reef, the reef forming a small bight or horseshoe bend…and in the darkness it was an awful sight to look at: the reef one sheet of white foam with points of black rock sticking out of the water."[154]

Fuller waited for better weather and tried to rig a makeshift rudder from his main boom, but the storm only increased in intensity. "I waited until noon, hoping it would moderate…instead of moderating it blew the harder and [with] more sea." By afternoon, he judged his best course of action was to take the crew in ship's boats to safety on shore before darkness arrived. There, they would wait out the storm and then return to *Pilot's Bride* if it was still there. They had to hurry as the shore was about eight miles distant and stormy weather hastened oncoming darkness. They landed at Betsey Cove and used *Trinity*'s spare material to erect a makeshift shelter.[155]

Storm conditions continued for the next forty-eight hours as crewmen anxiously tried to see the ship eight miles off. But murk and blowing snow concealed any views. Wind did moderate, but a heavy fog descended. "On [the] morning of the fourth day I was up bright and early as the wind had come around to the westward and went up to the top of the land and had a look off to the Rocks [of] Despair. I could see the rocks but no schooner. She had disappeared."[156]

They found *Pilot's Bride* wreckage scattered along the coast as far as Black Point. Men found Fuller's washstand had come ashore unscathed and along with it a large barrel of whiskey, which at first opportunity, they got into and became roaring drunk, so Fuller poured out rest of it. They also found the top of the ship's deckhouse. Within weeks, some crew members built themselves shanties from wreckage pieces. One group of Portuguese built a shack called Hotel de Portuguese. A mile farther on, others built a structure called American House. Crew cohesion fractured, and they argued and physically fought for next eleven months. Fuller retained some semblance of control but only with parts of his crew. They subsisted on stores salvaged from *Pilot's Bride* wreckage and anything *Trinity* had left. They also made ample use of local plants and seabirds. It was still a grim and desperate struggle. Finally, on September 11, 1883, they saw sails of a ship entering Norton's Harbor. It was *Francis Allyn*. When they got aboard, Fuller was too choked up to speak, and many stranded crewmen broke down and wept, as their ordeal was over. They also learned of *Trinity*'s loss at Heard Island. Fuller wrote that *Trinity*'s owners could not or would not send a vessel to look for them. Survivors were taken to Cape Town, and Fuller eventually returned to New London and later went back sealing to Kerguelen Island, this time as master of *Francis Allyn*, the ship that had rescued them. He wrote up his experiences and titled his work *Master of Desolation*. Captain Joseph J. Fuller died in 1920.[157]

*John Carver* was a 319-ton bark built in Searsport in 1857. Built by its namesake, John Carver, it made six voyages to the Arctic and northern Pacific starting in 1866. In 1869, it assisted in rescuing crew of the bark *Eagle*, which had grounded on Sea Horse Shoal in the Arctic. *John Carver*'s May 1880 voyage, under master Abram Smith, included Mainer William Gill of Portland. The crew returned in May 1883 with 830 barrels of oil, all but 130 of them sperm whale, and eight hundred pounds of baleen. Its final voyage was part of San Francisco's Shanghaier fleet. On June 21, 1886, while whaling twenty-five miles south of King Island in the Bering Sea, *John Carver* was crushed by ice. Its crew drifted in small open boats for over thirty hours. *John Carver*'s hulk later drifted to Cape Thompson, where its catch was saved. The Penobscot Marine Museum has a model of

Penobscot Marine Museum ship model of *John Carver* showing its whaling boats. *Photo by author.*

the locally built vessel, and well-known fellow Mainer Waldo Peirce, who also lived in town for a time, painted a scene of the ship on the stocks at its Searsport shipyard.[158]

In 1858, Searsport-built brig *A.J. Ross* was launched. It made three voyages to Hudson Bay starting in 1876 under master James N. Hyatt. Its 1877 voyage included Mainer Rufus H. Potter of Oxford and returned with 243 barrels of oil and 23,000 pounds of baleen. Portland's John Marley shipped aboard it in 1878 under James G. Sinclair. *A.J. Ross* was lost in Hudson Bay that year, and it is not clear whether Marley survived, but its voyage is credited with 20 barrels of whale oil. Hancock-built schooner *Astoria* made two voyages to the south Atlantic and totaled a combined catch of 194 barrels of oil and 350 pounds of baleen. It was condemned at Cape of Good Hope in 1880. Hancock-built, 92-ton schooner *D.N. Richards* made two voyages to Atlantic waters, starting in 1867, for a combined total of 312 barrels of oil, 202 of them sperm whale. It was condemned at Norfolk, Virginia, on September 20, 1870. Castine's schooner, *Eothen*, was built in 1860 and made two voyages from New York, with no results of any catch recorded. The 141-ton brig *Starlight*, which had been built in 1860 in Nova Scotia, was sold to Bangor owners in 1873 and did two whaling voyages to the south Atlantic that collected 2,159 barrels of oil, 1,375 of which were sperm whale.[159]

In 1862, Searsport bark *Emma F. Herriman* was built but was withdrawn from merchant service in 1864. It went into whaling in 1880 and operated out of San Francisco for fourteen voyages as a Shanghaier. Two voyages were under William T. Shorey, one of the industry's few black whaling captains. Shorey had worked his way up the ranks since 1876, when he signed aboard a whaler as a green hand. He proved himself a skilled whaleman and an excellent navigator. On his second voyage with *Emma F. Herriman*, he took it to north Japanese waters and the Sea of Okhotsk and returned with 570 barrels of oil and twenty-five tons of baleen. *Emma F. Herriman*'s career totals include 5,095 barrels of oil, 650 of them sperm whale, and nearly twenty-three tons of baleen. It was removed from whaling service in 1894.[160]

# 5. CIVIL WAR

*A ship in port is safe, but that's not what ships are built for.*
*—Grace Hopper*

The eruption of America's Civil War in 1861 dramatically affected the American whaling industry. Many Maine-built ships that had gone into whaling now found themselves involved in various aspects of the armed conflict. Some were attacked, captured and destroyed by Confederate raiders. Others, deemed beyond their usefulness, were purchased by the United States government and purposely sunk in Confederate harbors as part of what were called Stone Fleets. The Stone Fleet plan was initiated to prevent blockade runners from supplying Confederate interests by intentionally sinking old whalers in the mouths of Southern harbors. Some of the towed ships sank before even reaching their destinations. Others were sunk near Georgia's Tybee Island, to be used as breakwaters or wharves for landing Northern troops. Most of these condemned vessels were divided into two separate fleets, designated First and Second Stone Fleets. Filled with stone, sand and dirt, they were towed or sailed under their own power to their final destinations and scuttled. In 1861, twenty-four vessels of the First Stone Fleet were sunk in Charleston Harbor on December 19 and 20. In 1862, the Second Stone Fleet was sunk in Maffitt's Channel, near Charleston Harbor, in South Carolina. It is debatable whether this extensive effort was effective or not. One source refers to it as futile due to shifting sands and strong currents that changed the

The Stone Fleet in Charleston Harbor, December 19–21, 1861. *From* Leslie's Weekly Illustrated Newspaper.

shipping channels and made the obstacles irrelevant. Confederate general Robert E. Lee took a dim view of the effort and called the attempt "an abortive expression of the malice and revenge" of the North.[161]

In December 1861, writer Herman Melville, of *Moby-Dick* fame, was moved enough by the event to write a poem about the senselessness of their fate:

*The Stone Fleet: An Old Sailor's Lament*

*I have a feeling for those ships,*
*Each worn and ancient one,*
*With great bluff bows, and broad in beam;*
*Ay, it was unkindly done.*
*But so they serve the Obsolete—*
*Even so, Stone Fleet!*
*You'll say I'm doting; do but think*
*I scudded round the Horn in one—*
*The* Tenedos, *a glorious*
*Good old craft as ever run—*
*Sunk (how all unmeet!)*

*With the Old Stone Fleet.*
*An India ship of fame was she,*
*Spices and shawls and fans she bore;*
*A whaler when her wrinkles came—*
*Turned off! Till, spent and poor,*
*Her bones were sold (escheat)!*
*Ah! Stone Fleet.*
*Four were erst patrician keels*
*(Names attest what families be),*
*The* Kensington, *and* Richmond *too,*
Leonidas, *and* Lee*:*
*But now they have their seat*
*With the Old Stone Fleet.*
*To scuttle them—a pirate deed—*
*Sack them, and dismast;*
*They sunk so slow, they died so hard,*
*But gurgling dropped at last.*
*Their ghosts in gales repeat*
*Woe's us, Stone Fleet!*
*And all for naught. The waters pass—*
*Currents will have their way;*
*Nature is nobody's ally; 'tis well;*
*The harbor is bettered—will stay.*
*A failure, and complete,*
*Was your Old Stone Fleet.*[162]

To ready them for scuttling, a two-inch hole was bored in the counter above the light-water line, which ran through the vessel. These were then plugged and bolted together. At the proper time, the bolts would be knocked out, making the plugs fall away and allowing water to flow in. New Bedford resident James Duddy supplied the stone for this fleet. He soon had most area farmers tearing down their stone walls and loading them on sleds and wagons. Crews were engaged for their final voyages to Charleston. The captains elected Rodney French as commodore, and he was later mayor of New Bedford. The vessels gathered in Charleston's channel on December 19 and 20, and acting master George H. Bradbury positioned each ship, knocked their bolts out and sank the ships to the bottom.[163]

South Berwick brig *Leonidas,* built in 1833, was bought on October 27, 1861, by the U.S. Navy for $3,050. Slated for the First Stone Fleet, it was

Charleston Harbor and site of Stone Fleet wrecks. *From* Harper's Weekly, *January 11, 1862.*

filled with 200 tons of sand and stone and sailed from New Bedford under master John Howland on November 20, 1861. It was intentionally sunk one month later, along with fifteen other old whalers, four miles south-southeast of Fort Sumter and three miles east-southeast of navigation light on Morris Island. *Leonidas* was immortalized in Melville's poem. Nobleboro's 254-ton bark *Cossack*, built in 1835, was also purchased by the United States Navy. Filled with 250 tons of stone and sand, it was towed south and beached on Georgia's Tybee Island on December 8, 1861, and acted as a wharf for landing Union troops. Another was bark *Noble*, built in 1831. After being condemned at Auckland in 1846, *Noble* entered merchant service and then became part of Second Stone Fleet. It was intentionally sunk in Maffitt's Channel. Brunswick brig *Edward*, built in 1832, may have been part of the Second Stone Fleet and was intentionally sunk in Maffitt's Channel in 1862. Schooner *Mac*, built in Thomaston in 1831, also became part of the Second Stone Fleet, as was *Majestic*, built in Thomaston in 1828. Both were sunk in Maffitt's Channel in January 1862.[164]

One Civil War casualty not attributed to Stone Fleets or Confederate raiders was whaler *Polar Star*. Built in Mattapoisett, Massachusetts, in 1852, it was lost in May 1861 in the Sea of Ohhotsk. Its crew, under master Daniel D. Wood, included forty-four-year-old Mainer Henry J. Durgin of Brownfield. It is unclear if he survived. Sanford natives John Ford and John Smith had been part of *Polar Star*'s August 1856 whaling crew under master Hiram Weeks. They returned in July 1860 with 2,300 barrels of oil, 558 of which were sperm whale, and over nine tons of baleen.[165]

Many whaling ships fell victim to Confederate raiders, some of which had been built in Maine or had Mainers on board. On December 8, 1861, CSS *Sumter* captured whaler *Ebenezer Dodge* out of New Bedford en route to Pacific whaling grounds. The whaler's provisions, water, clothing, small stores and a dozen whale boats were confiscated by *Sumter*. Known as *Eben Dodge*, its November 1861 voyage had begun under master Gideon C. Hoxie and had included Mainer Jacob W. Hyman of Isleboro. An earlier 1856 voyage of *Ebenezer Dodge* had included Mainers Joshua F.L. Dennett of Hollis and William Lewis of Augusta. Its 1858 voyage included John Riley of Eastport. In December 1861, after its capture, *Eben Dodge* was put to the torch.[166]

On September 8, 1862, CSS *Alabama* encountered 313-ton whaling bark *Ocean Rover*. A New Bedford whaler with a crew of thirty-six, it was at N 39°50′, W 35°25′45″ with a cargo of 1,100 barrels of oil when captured. Mainer Amos Handy of Bath was aboard—he first sailed the ship in 1859. It was burned on September 9. *Ocean Rover*'s crew took to six whale boats and rowed to the Azores. Less than a week later, on September 16, *Alabama* captured 257-ton schooner *Courser* out of Provincetown and then used it for target practice. *Courser* burned that forenoon and sank five miles from Flores, Azores. Mainers Herbert Haskell of Deer Isle and Joseph Hill of Wells were aboard. Records indicate that Haskell returned, but it is unclear about Hill. Boston-built *Alert*'s 1862 voyage, under Edwin Church, to Kerguelen, Desolation and Heard Islands for elephant seals was able to send back 220 barrels of oil, before it too was caught and burned by *Alabama*. Mainers aboard were Francis Stevens and Eastport native David Gelbon. In November 1862, *Alabama* cruised near Bermuda and came on 376-ton whaler *Levi Starbuck* out of New Bedford. It was headed for Pacific whaling grounds. On November 2, *Levi Starbuck* was captured and burned after its stores had been removed. Mainer Stephen Trafton of Machias was aboard, and he eventually returned to the United States. *Levi Starbuck*'s 1857 voyage also included Mainers William F. Shanatt of Portland and Edward H. Wilson of Waldoboro, who may still have been aboard. By April 1863, *Alabama* was off Brazil when it came on 132-ton whaling brig *Kate Cory* out of New Bedford. On April 15 at Fernando de Noronha, *Kate Cory* was captured four miles from land. Mainers William Gaven of Bangor and Joseph Chamberlain of Portland were aboard. The old whaling bark *Lafayette*, out of Westport, Connecticut, was captured by *Alabama* the same day as *Kate Cory* and had aboard Mainer John G. Watts of Yarmouth. Within a week, *Alabama* also captured 211-ton whaling bark *Nye*. With a crew of twenty-four, *Nye* carried 500 barrels of oil, 425 of them sperm whale. On April 24, it was put to the torch. Mainers Rueben O. Eldred of Pittston and Rufus A. Whittier of Belfast were aboard.[167]

One of the most successful Confederate raiders was CSS *Shenandoah*, which sank sixty-four American ships. On December 4, 1864, the old whaling bark *Edward* was captured and burned off Tristan de Cunha. Its 1846 voyage included Stephen Barnum of Alfred, Maine. William N. Richards of Greenwood, Maine, sailed on it in 1856. The following April 3, whaling bark *Pearl* of New London was captured and burned at Lodh Pah Harbor on Pohnpei Island in Micronesia. *Pearl*'s 1852 voyage included Alonzo W. Creany of Bath. Amos Maxfield of Randolph, Maine, sailed it in 1854. On April 4, 1865, *Shenandoah* captured and burned whalers *Hector* of New Bedford and *Edward Carey* of San Francisco at Lodh Pah Harbor. Hector's 1856 voyage included Mainers Amos A. Chase of East Limington and William M. Ray of Bath. Mainer Alvin S. Warren of Lovell was aboard *Hector*'s 1861 voyage. Less than a week after burning *Hector* and *Edward Carey*, whaling bark *Harvest* was put to the torch at Lodh Pah Harbor after supplying *Shenandoah* with Arctic charts for its next attack in the Bering Sea.[168]

By May 1865, *Shenandoah* was in the Sea of Okhotsk, where on May 27, it captured and burned whaling bark *Abigail* of New Bedford. *Abigail* carried a large quantity of whale oil and liquor. Mainer John C. Nights of Lincoln was one of *Abigail*'s thirty-five crew members aboard, fourteen of whom joined *Shenandoah*. It is not clear if Nights did so. Several crew members did get drunk from whiskey that was seized. At least three Maine-built whalers fell afoul of Confederate raider *Shenandoah* when it made its way into the Bering Sea with the help of nautical charts taken from whaler *Harvest* while in the Pacific. Throughout the spring and summer of 1865, it spread havoc among the American whaling fleet and destroyed twenty of fifty-eight Yankee whalers in the Bering Sea. Ironically, most destruction occurred after Confederate general Robert E. Lee had surrendered at Appomattox Court House in April 1865.[169]

In the third week of June 1865, *Sophia Thornton* was at N 62°40′, W 178°50′ when it was captured and burned by *Shenandoah*. Georgetown's 1851-built bark was brought to with a warning shot across its bow on June 22, 1865. Captured men were ordered to take what provisions they wanted and get into boats. They were ordered to set fire to the whaler. To keep them from simply sailing away after *Shenandoah* left, *Sophia Thornton*'s masts were cut down.[170] *Shenandoah*'s first mate, Cornelius Hunt, wrote, "We could see the disabled vessel, with her masts dragging alongside, and the paroled prisoners in their whale boats, transferring from her to the Milo whatever suited their fancy. I have no doubt the craft was thoroughly ransacked, but ere the sun made its brief disappearance below the horizon, a bright tongue of flame shot heavenward, telling us the prisoners had performed their distasteful task."[171]

That same day, *Shenandoah* captured 364-ton whaling bark *Euphrates*. After removing its navigation instruments, *Euphrates* was torched. That old whaler's 1849 voyage included Maine brothers James and Peter Treanor of Cherryfield and Henry Wragg of Peru. Thomas Higgins of Portland sailed it in 1851. The whaling bark 428-ton *Jireh Swift* was also captured and burned just after *Euphrates*. Mainers aboard *Jireh Swift* included Robert Curtis of Monroe and Alfred Davis of Eastport. One captured whaler—the oldest and most decrepit of the fleet—escaped destruction. *Milo* had been built in 1811 in Newburyport, Massachusetts. Mainers Granville W. Robinson of Vassalboro was the first mate on its 1849 voyage. Now, under master Jonathan Capen Hawes, it included Mainers Washington Ney of

Greene and Edward Irving of Vassalboro. They told the Confederates that the war was over but were not believed. *Shenandoah*'s officers bonded *Milo* for $46,000 to transport the captured whalemen from other ships. *Milo* was sold in 1872.[172]

Still on June 22, *Shenandoah* captured and burned 159-ton brigantine *Susan and Abigail*. New Bedford-built, 495-ton *William Thompson* was next. Northeast of Cape Narrows, Shenandoah captured and burned the ship. Donald McPartine of Bangor was aboard. The old whaler had seen lots of service. Its 1842 voyage included Mainers James Hathaway of Bath and William Rimick of Eliot. Wyndham R. Bates of Portland and William Lawson of Eastport sailed it in 1846. Its 1857 voyage included Portland men Thomas Fields, William H. Jones and Frank B. Shaw. O.D. Lovejoy of Bangor was also aboard then. Now, *William Thompson*, *Abigail*, *Jireh Swift*, *Euphrates* and *Sophia Thornton* were smoking hulks slipping below the surface of the Bering Sea. *Shenandoah* was not finished with its rampage. On June 25, 1865, 419-ton *General Williams* was near St. Lawrence Island in the Bering Strait when *Shenandoah* captured and burned it. Eastport native Angus O. Henley was one of the crew members. The old whaler's 1845 voyage included Mainers Joseph Smith, George Eascott of Lubec, Henry Williams of Portland and Patrick Buskby and Dennis Moran, both of Eastport.[173]

The following day, June 26, *Shenandoah* captured 340-ton *Nimrod*. Its captain, James Clark, had lost his previous ship, *Ocean Rover*, to *Alabama* near the Azores. Richard Walls of Scotland, Maine, sailed aboard *Nimrod* in 1857, but Amos Handy of Bath was aboard when it was taken. *Shenandoah* next captured and burned 1844-built Bath bark *Isabella*. *Catharine*, a 389-ton whaling bark with a nearly full cargo of oil, was also taken and burned, followed by whaling ship *Gipsey* (or *Gypsy*). This was the same *Gypsy* whose 1856 voyage included Mainers Ansell W. Durfee, Hagen Lewis of Fryeburg and John C. Oliver of Bath. Robert Phelon of Eastport sailed *Gypsy* in 1857. Then it was 389-ton whaling ship *William C. Nye*'s turn. Its 1857 voyage included Mainers Rowland C. Clapp of Sedgwick, Franklin W. Harradan of Turner and possible brother Harvey Harradan of North Auburn, H. F. Smith of Clinton and Samuel H. Wentworth of New Portland. That same day, June 26, *Shenandoah* captured *General Pike*. Instead of being put to the torch, *General Pike* was sent to San Francisco with over two hundred prisoners aboard under $30,000 bond. When its captain complained to the Confederates about overcrowding and lack of provisions, *Shenandoah*'s captain told him to "eat the Kanakas [Hawaiian crewmen] if you get hungry." *General Pike* spent its final eight whaling expeditions in the Arctic before being condemned and sold in Tahiti in 1868.[174]

June 28, 1865, proved particularly destructive for America's whaling fleet. *Shenandoah* moved at will, capturing and burning whaling bark *Congress*. George Simms of Eastport sailed *Congress* in 1842 out of Mystic. Now, this was *Congress* master Daniel Wood's third lost vessel in four years. Maine native captain Thomas G. Young's *Favorite* was also taken and burned. Mainer George Lovell of Parsonsfield had shipped her in 1857 and may still have been aboard. On June 28, *Hillman* was also taken and burned. An earlier 1857 voyage had included Mainers Charles Sawyer of Surry, George Sharpless of Eastport and James Edwin of Ellsworth. *Martha* also fell to *Shenandoah* on June 28 in East Cape Bay. That old whaler had seen much service. Mainers Michael Power of Eastport and William H. Getchell of Sebec sailed it in 1841, William Shaw of Bangor sailed it in 1845 and *Martha*'s 1858 voyage had included Mainer George W. Greene of Limington. *Nassau*, built in 1825, was also captured and burned by *Shenandoah* on June 28. Its 1856 voyage included Mainers James S. Haskins of Kennebec and Joel Simpson of Newcastle. *Nassau*'s 1860 voyage included Joseph S. Chamberlain and James McTier, both of Portland; Ephram Horne of North Anson; greenhand Joshua Winslow of Mercer; and John Blake of Denmark.[175]

*Waverly* was yet another whaler captured and burned by *Shenandoah* on June 28. It was caught in dead calm waters near Diomede Islands. Its crew of thirty-three included Barnard Edwards of Rockland, Maine, who did return from the ordeal, as well as James S. Holt of Oxford. *Waverly*'s 1846 voyage had included Isaac E. Farrald of Hallowell. Its 1859 voyage included James Colliers of New Sharon, Maine. *Brunswick*, built in Brunswick in 1827, was also captured and burned in the Bering Strait in June 1865 by *Shenandoah*. On June 27, it had collided with a large cake of ice and stove in its hull twenty inches below starboard bow waterline. It listed severely, and captain and crew made plans to abandon ship. The following day, its rigging, gear and anything else salvageable were put up for immediate auction to nine other whale ships anchored in the area. *Shenandoah*'s arrival put an end to any sale, and *Brunswick* was burned. In the final military action of the conflict, *Brunswick* was the final maritime casualty of the American Civil War.[176]

As June 28 drew to a close, CSS *Shenandoah* captured two other whaling ships but used them to transport captured crewmen back to civilization. *James Maury* of New Bedford was bonded for $37,600 in East Cape Bay, while 1826 New York–built whaling bark *Nile* bonded for $41,000. Both were loaded with prisoners. *Nile* was sent with 222 men to San Francisco. *James*

*Maury*, under master Slumon L. Gray, eventually landed a cargo of 261 barrels of sperm whale oil, 1,656 barrels of regular whale oil and 26,333 pounds of baleen. Aboard at the time were brothers Franklin W. and Harvey Harradon Jr. of Turner, Maine. As early as 1848, Charles Brown of Portland had shipped aboard *James Maury*. William Taylor of Bangor and Dennett F. Beane of Newport were part of its 1855 crew, while Reuel Cunningham of Litchfield, George N. Adams and William Brinizine Jr. of Monmouth sailed in 1859. *James Maury* went on one final whaling voyage in 1868 under master John C. Smith, in which she netted 1,830 barrels of sperm whale oil, 1,038 barrels of regular whale oil and 500 pounds of baleen. It was later sold to New York owners in 1873 and then to foreign owners. *Nile* made five more whaling voyages. Mainer Frank Hopkins of Kittery shipped aboard *Nile* to the north Atlantic and Cumberland Inlet in 1874 under master John Orrin Spicer and came back with 800 barrels of oil and 4 tons of baleen. G.H. Pomeroy of Northport went north with *Nile* in 1875, again under Spicer, and returned with 380 barrels of oil and 2.5 tons of baleen. *Nile* was then converted to a barge at New London, Connecticut.[177]

The Civil War seriously affected the whaling industry. The loss of so many vessels and a slowdown in construction weakened the fleet. Many older vessels were destroyed in Stone Fleet operations, while others were captured and sunk by Confederate raiders. These challenges, coupled with dwindling stocks of whales, did not prove promising for the future of whaling. Those whalers and whalemen who did survive the war now found whaling grounds more elusive than ever. The drastic slowdown in Maine ship construction began to change after hostilities ceased in 1865. Maine shipbuilders began producing new vessels with new technology, and Maine men once again answered the call and went whaling. But even with technological advancements and newer ships, American whaling had entered its twilight years, and the end of an era had begun.

6.

# END OF AN ERA

*There is nothing more enticing, disenchanting, and enslaving than the life at sea.*
*—Joseph Conrad*

Kennebunk produced numerous schooners at war's end, and some made their way into whaling, including 117-ton *William A. Grozier*. Built in 1865, it made nearly forty voyages to the Atlantic and Indian Oceans, starting in 1866, until it was put into Cape Verde packet trade in 1911 and was lost in the mid-Atlantic in 1914. Its whaling career total included an astonishing 14,958 barrels of oil, all but 566 of them from sperm whales. Kennebunk-built, 118-ton schooner *J. Taylor* made two whaling voyages out of Provincetown, Massachusetts, from 1866 to 1869, both under master Atkins Smith. It returned with a combined catch of 405 barrels of oil, all but 180 of them sperm whale. Schooner *Charles Thompson* made six voyages to the Atlantic starting in 1874, most under master Whipple A. Leach. Its career totals include 1,982 barrels of oil, 1,844 of which were sperm whale. It was lost in August 1879. Kennebunk-built 116-ton schooner *Benjamin T. Crocker* (*B.T. Crocker*) made one voyage out of Provincetown to Atlantic waters under master John S. Chandler in 1866. They returned in July 1867 with 157 barrels of oil, forty of which were sperm whale. It was withdrawn from service in 1868.[178]

In 1866, Kennebunk schooner *Thomas Hunt* was built. It made two voyages to the south Atlantic, starting in 1884. Its second voyage, under Jose Antonio Santos, returned with 1,300 barrels of oil, 850 of which were sperm whale. It was lost off the Cape Verde Islands in 1890. Kennebunk-built schooner *John Atwood* made one whaling voyage out of Provincetown to Hudson Bay in 1872 under master Elnathan B. Fisher. It returned with only 180 barrels of oil and 3,128 pounds of baleen. It was withdrawn from service later that season. Kennebunk-built schooner *Willie Irving* made one whaling voyage to Atlantic waters in 1867 under master Richard S. White and returned with 160 barrels of sperm whale oil. It is thought that it foundered near George's Bank later that year. Gouldsboro-built brig *Sullivan* made six whaling voyages to the Atlantic starting in 1905, until it was lost at Horta Bay at Fayal on October 14, 1913. It is not listed in the NMDL database. Searsport-built bark *Augustine Kobbe* was 136 feet long and 532 tons. It appears to have made one whaling voyage out of New Bedford in 1894 under master Henrique S. Oliveira. Another record states the name was H.S. Oloverro. There are no records of any catches. It was reported that it was sold to foreign owners in 1901.[179]

In 1867, 180-ton schooner *John H. Perry* was built in Bath but is not listed in NMDL's database of American whaling vessels. Bath residents also built schooner *Delia Hodgkins* in 1870. It made three whaling voyages

Bark *Augustine Kobbe*, built in 1866 in Searsport, Maine. *Penobscot Marine Museum.*

to the north Atlantic, Cumberland Inlet and Hudson Bay starting in 1879, all under master Stanford Stoddard Miner. Its career total haul was eight hundred barrels, five hundred of which were sperm whale. It capsized off Cape Cod on November 4, 1881. There was also Bath-built schooner *Mary E. Higgins.* In a rare occurrence, *Delia* and *Mary E. Higgins* were launched the same day at Hodgkins & Hawthorn shipyard. They had nearly identical dimensions, both being 127 tons and built for Union Wharf Co. of Provincetown, Massachusetts, for fishing and oyster trade. *Marry E. Higgins* made two disappointing whaling voyages out of New London from June 1879 to April 1882 under master Benjamin Nelson Rogers. It returned empty both times.[180]

As whale population numbers decreased, whaling ships went farther in search of them. Arctic waters were resource-rich, especially north of the Bering Sea off Alaska. American whale ships had poked into Bering Sea waters since 1848. In that year, a New York whaler reported numerous bowhead whales. It was not long before ships ventured farther north of the Bering Strait into waters off Point Barrow, Alaska. But that proved a difficult journey, as savage Arctic conditions provided numerous challenges. Timing was critical; whalers had to be on station by midsummer, when ice finally broke up and permitted limited passage in treacherous waters. Whale ships and crews were done and out by late summer to avoid getting caught in new ice. Quickly formed ice could easily trap and crush vessels. Whaleships often wintered at Hawaii. By the 1880s, whaling captains opted to freeze themselves in for winter to stay longer on station and get in two seasons' worth of whaling. A favorite winter anchorage became Herschel Island.[181]

Wintering in Arctic ice was very dangerous, and a series of Arctic disasters struck American whaling shortly after the Civil War. Other factors challenged American whaling as well. By 1870, there were noticeably fewer whales and a growing demand for competitive fuels such as kerosene and oil. Rising expenses for outfitting and crewing whale ships also hurt the business. The *New York Times* reported the state of whale fishery in 1870 as one of poor returns and declining prices, lower than any since the Civil War. "Our merchants do not look upon the future of whaling with encouragement, and seem disposed to distrust it as to its pecuniary results, induced more by extraneous causes than inherent." Predicted decline in the number of whaling fleet vessels were realized when the *New York Times* reported only 216 ships and barks, 18 brigs and 54 schooners were ready for the 1871 season.[182]

Right whaling in the Bering Straits and Arctic Ocean with its varieties. *Library of Congress.*

As early as April 1871, whaleships began setting courses for the Bering Sea, often with a Honolulu stopover to unload whale oil and whalebone already accumulated. During that 1871 summer in Alaskan waters, bark *Florence*, built in Bath in 1840, came on *Minerva*, which had been caught in ice and abandoned off Wainwright Inlet. The crew salvaged *Minerva* and brought it back to San Francisco, where it was sold. *Minerva* was the initial loss of that 1871 season. What followed was a series of disasters that rocked American whaling. By June, ice was open as far north as Cape Navarino. American whale ships worked their way to Cape Bering and Flower Bay, although they encountered thick ice. By September, ice floes and bergs drifted down from the north, and by September 10, a number of vessels had been sunk or were hemmed in by drifting ice. Some were driven ashore.[183]

The *Friend*, a Hawaiian newspaper, reported on November 1, 1871, that thirty-three whale ships were trapped in thick ice off northern Alaska. Another source put the number at thirty-two.[184] The *Friend* article stated that whaling captains have reported their "ships cannot be got out this year, and there being no harbor that we can get our vessels into, and not having provisions enough to feed our crews to exceed three months, and being in a barren country where there is neither food or fuel to be obtained, we feel ourselves under the painful necessity of abandoning our vessels, and trying to work our way south with our boats, and if possible to get on board of ships that are south of the ice."[185]

The *New York Times* followed on November 6 with a short article of the disaster and a more detailed description the next day. It was now mid-September, and whaling captains were desperate. Many felt that they had no chance, since early winter storms would certainly grind ice together and crush their ships. They wrote if they "be cast upon the beach it would be at least eleven months before we could look for assistance, and in all probability nine out of ten would die of starvation or scurvy before the opening of spring."[186]

One of nineteen whale ships out of New Bedford that was lost that season off Alaska was 368-ton, Freeport-built *Emily Morgan*, although the *New York Times* recorded it as *G. Morgan*. Its forty-year whaling career was in jeopardy, and its 1871 season was *Emily Morgan*'s twelfth whaling voyage and sixth in a row to Arctic waters. Earlier that season, it had come on a beached wreck of whaling bark *Oriole* at Plover Bay. For $1,350, *Emily Morgan*'s captain Benjamin Dexter bought *Oriole*'s remains as is. He stripped it and resold much of the gear and equipment to other nearby whalers for a net profit of $1,200. *Oriole*'s 1857 whaling voyage included Melvin C. Bachelder of Hallowell, Maine. Its 1870 voyage, which ended with the ship being wrecked on the beach, included Mainer Alton O'Brien of Buckfield. While forty-one other whalers worked their way north in May 1871, *Oriole* instead followed the coast to within a few miles of Icy Cape and reached open water near Plover Bay with no other ship in sight. A large cake of ice kept the ship from passing it to windward. It veered near enough to the ice that an underwater ice projection sliced into the ship below its waterline. By evening, water in *Oriole*'s hold was up to the deck beams. Pumps were manned and the hole was plugged with oakum and canvas. Beaching on shore at Plover Bay, the crew repaired damage, but the ship's after-hatch broke under pressure from the water, and *Oriole* sank in three minutes with only its mastheads out of the sea.[187]

Now *Emily Morgan* was in danger of getting caught in ice. In a five-day period before being beset, it had tough luck catching whales when "she lowered and struck a whale with two boats, losing him when one of the lines parted and an iron drew; sent five boats into the ice, struck another whale, losing him and the line as well; saw a number of whales, struck one, and lost him; and sent boats back into the ice in a heavy fall of snow, hunted for hours, and saw nothing, the crew returning to the ship weary and half-frozen." Complaints naturally arose and the mate recorded that one boat-steerer was ordered to keep silent as he "was having too much to say." When he refused to stay silent, his discipline was having his jaw broken.[188]

By August 29, the trap had closed and shifting ice and fog conditions pinned in the whalers. On September 7, *Emily Morgan* spotted whales and struck one, but second mate Antonio Oliver was killed when his bomb gun accidently went off. "The bomb entered midway of his lower jaw near the throat and came out at the back of his head, killing him instantly."[189] The crew buried him on shore in a half-frozen hole while the mate read from the Bible. By September 9, *Emily Morgan* was one of seven whalers stranded near Icy Cape, all in considerable danger. Crews began making bread bags for supplies in case they abandoned ship. But immediately taking to whaleboats was considered extremely foolish by *Emily Morgan*'s captain Dexter. He stated that he "will not cross the Arctic Ocean in an open whale-boat laden with men and provisions in the latter part of the month of September and October. As far as Icy Cape there is no danger, but beyond that, the sea is dangerous at this season of the year. Out of the fourteen hundred men not a hundred will survive."[190]

But its crew would have none of it and planned to abandon *Emily Morgan* off Point Belcher on September 14, 1871. Dexter wrote, "All hope of saving this ship or any of the others has entirely vanished. We can only expect to escape with our lives, if even that."[191]

One evacuee from *Emily Morgan* was the captain's wife, Almira Dexter. Her last log entry states:

> *All hope of saving ship or property is gone. If we save our lives, we ought to be satisfied and that should satisfy the world. To winter here might be possible, but not under these circumstances….At twelve noon, paid out all of chain on both anchors and at 1:30 PM, with sad hearts, ordered all the men into the boats and with a last look over the decks, abandoned the ship to the mercy of the elements. And so ends this day, the writer having done his duty and believes every man to have done the same.*[192]

*Emily Morgan*'s whaleboats worked steadily southward in a driving rain throughout the night. Its first mate later recorded, "As night came on, the wind increased and as darkness closed around us, heavy black clouds seem to rest over us and it was not possible to see more than a few feet and we were in constant danger of coming in collision with the many fragments of ice floating in the narrow passage between land and the main pack." By early morning, they found themselves still over twenty miles from Icy

Cape, which they finally reached that afternoon. Next, they had to cross open water to rescue ships anchored offshore. Battling freezing spray and constantly bailing, they finally reached *Europa* more than twenty-four hours after they had abandoned *Emily Morgan*. The whaler was later found to have such a large hole in it that tides ebbed and flowed right through it. The loss of ship was valued at $60,000. *Europa*'s Maine connection was that its earlier 1862 voyage had included John Grant of Frankfort. Another Maine connection with this disaster was Captain Leander C. Owen. He was the first-time master of whaler *Contest*, another of the stranded vessels that was abandoned. Owen went on to captain other whale ships and was involved in the Arctic disaster of 1876.[193]

In all, 1,200 men and several women and children abandoned the whaleships and piled into smaller ship boats. As they proceeded south, captains saw the full extent of this disaster. For miles, there was nothing but solid ice holding the unfortunate whalers. A small seam of shoal water near a barren coast was the only avenue of escape. For seventy miles, men struggled along and slept on the beach, trying to avoid biting winds until they reached Icy Cape, where they encountered seven ships in open water just south of the ice. In rising winds and risking their own ships, these seven vessels got the crews aboard and put in at Plover Bay to get enough provisions to get them safely back to Hawaii. By the spring of 1872, only one stranded ship remained usable. The rest had either sunk, burned or washed ashore to be stripped for use by local inhabitants.[194]

*Florence* continued to make itself useful when it salvaged oil and bone aboard some 1871 abandoned vessels. It returned in the fall of 1872, with news of abandonment and the loss of 215-ton bark *Helen Snow*, which had been built in Bath in 1851. That 1872 season, *Helen Snow* returned to salvage wrecks from 1871's disaster. On August 19, while at anchor, it was surrounded by ice, which dragged it north into the pack. Against captain's orders, the crew abandoned the ship and made its way to nearby whalers. The pack ice relented, and much to chagrin of *Helen Snow*'s crew, the ship survived unharmed deep in the icepack. It was later spotted by Captain Owen, who got a salvage crew aboard, worked it free of ice and brought it back to San Francisco for sale. *Helen Snow* returned to Arctic ice fields as Hawaiian bark *Desmond* and was trapped once more. Again, *Helen Snow* was salvaged—this time by Russians who renamed it *Tugar*.[195]

The 1871 loss of thirty-three whaling ships was devastating to crews, owners, insurers and the whaling industry in general. In Hawaii, local papers reported that business came to a standstill with the loss of these Arctic whale ships. Many idle crewmen worked on sugar plantations, while owners and captains tried to recoup their losses, mostly through legal channels. Many lawsuits for reimbursement or salvage did not get settled for years. As late as April 19, 1880, the *New York Times* noted costs of rescuing whalers, such as thirty-five dollars in gold passage fee for each crewman. The newspaper also lamented that eight years had lapsed with no resolution about reimbursement of rescue vessels for their inability to catch whales due to being involved with the rescue. In fact, "their owners asked the Government to make good the loss occasioned by the interruption of their work, and the Treasury Department, in 1872, turned the case over to Congress for settlement."[196]

The 1871 Arctic disaster was such a setback that the *New York Times* devoted numerous articles and follow-ups about it, including a twenty-year retrospective. Even scrimshaw carvers were inspired to commemorate it. As late as 1998, there was proposal for the site of the disaster to be recognized as a historic landmark, especially once scientists and archaeologists precisely located the wreckage of the whaling ships. In 2003, maritime historian Randy Beebe found that an archaeological investigation had been done on a house that had been built using salvaged materials. He located and documented the disaster site during three seasons of fieldwork. Material culture found so far includes large hull sections, rigging hardware, copper sheathing, barrel hoops and staves.[197]

By 1874, newspapers gloomily reported prospects about low numbers of whales caught. Another crippling freeze struck Arctic whaling in 1876 as a dozen whale ships were once again caught in ice and lost. Again, bark *Florence* brought back news of the disaster when it returned to San Francisco with several survivors on October 21, 1876. Other survivors had gone aboard bark *Three Brothers* bound for Honolulu. Like before, whaleships had again entered Bering Sea waters in July and worked their way toward Point Barrow. While few whales were taken, crews did find a decent number of walruses, which filled oil casks. By August 14, fourteen ships worked off Point Barrow. Winds shifted two days later, and ice closed in, forcing ships to run as far south as Cape Smith. They were only able to maneuver in a tight swath of water, which stretched eight miles from shore as the open stretch steadily closed. By August 23, all ships were hemmed in by ice. *Florence* managed to drift up to Cape Smith and luckily got under lee of a grounded iceberg, which proved its salvation. Crews from other ships north of *Florence* began retreating south, hauling sledges to safety. Exhausted men finally reached beset *Florence*. Rather than continue their treacherous land journey, the men decided to try to winter at Cape Smith. On September 13, ice

Lithograph of Currier & Ives's Whale Fishery: Attacking a Right Whale. *Library of Congress.*

unexpectedly broke up before a strong east wind, and *Florence* was able to get underway. It made it to Sea Horse Island by evening and twenty-four hours later was off Wainwright Inlet. There, it waited for other crews, who finally decided to abandon their vessels. Two other ships also managed to extricate themselves and shared added crewmen from ships unable to escape.[198] "It is the undivided opinion of every master that no hope can be obtained of the rescue of the ships or those who remained on board of them. All are undoubtedly lost and carried away to the north-east in the immense ice-pack which closed them in for miles around."[199]

It was another blow to an already reeling industry. Insurance costs totaled nearly $500 million—many vessels had been insured for $30,000 to $40,000. This did not include losses from oil and bone taken. Still, a few New Bedford owners thought losses were not nearly bad as those suffered in 1871, mostly due to better insurance protection.[200]

New technology was being developed to counteract these losses. In the 1870s, Bath, Maine, led the way with a new type of whaler called steam barks. These were reinforced wooden ships with auxiliary steam engines and had been patterned on Dundee whalers that were used in Greenland and

Newfoundland waters. New Bedford whaling captain William Lewis placed an order for one with the Bath firm Goss, Sawyer and Packard in 1879.[201] However, before *Mary and Helen* was launched, Maine still produced more traditional ships that became whalers.

*Fresno* was a 160-foot, 1,250-ton schooner built in Bath in 1874 by W. Rogers. Its sailing on the Kennebec River was noted by a daily newspaper: "Sailed this morning the bark Fresno, 1,250 tons, recently launched by William Rogers, for New York in charge of a river pilot, a fresh breeze blowing from the northwest. In these days of steam the sailing of a large vessel under canvas from this port is a very rare occurrence."[202]

*Fresno* engaged in Alaska passage work, carrying whaling supplies and oil to and from Seattle. In 1922, after many years of service, it was brought to Lake Washington and moored alongside the wharf at Meydenbauer Bay in Bellevue. There, it was equipped with steam winches as wharf-side support for whaling operations. Within the year, fire broke out in its oil-soaked hold, and by the time a nearby tug and three whaling steamers had pulled it from the wharf, it was a total loss. Scrap firm Nieder & Marcus purchased *Fresno* for $1 but backed out when it found out the ship had settled on the bottom of the bay. Its owners, North Pacific Sea Products, raised and towed it north to be scuttled. After being stripped of usable machinery, *Fresno* was scuttled in 190 feet of water north of Evergreen Point Bridge, an effort that cost over $2,400, which prompted them to sue Neider & Marcus. Eighty years later, *Fresno* was rediscovered as one of three mystery wrecks Washington State Department of Transportation (WSDOT) thought might be situated in the way of anchor cables for a proposed replacement bridge.[203]

Another 1874-built Maine vessel that found its way into whaling was Columbia Falls ship *Leonora*. It made two whaling voyages out of New Bedford to the south Atlantic and Desolation Islands from 1901 to 1906 under master Benjamin D. Cleveland and then William B. Ellis. One of those voyages included twenty-two-year-old John Pierce of Portland. Cleveland brought back 2,900 barrels of whale oil, while Ellis brought back 2,440 barrels of oil, 2,290 of which were sperm whale. Bath schooner *Adelia Chase* was built in 1875 and made nineteen whaling voyages out of New Bedford to Atlantic waters, starting in 1880. That first voyage lasted two years under Captain Erastus Church and visited the Falkland Islands and the South Shetland Islands, where the crew took fifty-seven fur seal pelts. Crew members worked the South Orkney Islands and the South Sandwich Islands. In 1882, they sighted Bouvet Island but did not land. *Adelia Chase* returned with 175 barrels of oil, 15 of which came from one sperm whale.

It was more successful on voyages in 1890 to 1893 and 1902 to 1904. The first of the two was the most productive, with 1,350 barrels of oil brought back by Captain Michael A. Ferriera. The ship's 1902 voyage under master Ayres J. Senna brought back 1,240 barrels of oil. *Adelia Chase* was no longer whaling by 1906, when it was abandoned at Brava, Cape Verde Islands.[204]

*John and Winthrop*, a 338-ton Bath bark built in 1876, made twenty-seven whaling voyages to the north Pacific, the Sea of Okhotsk and Japanese waters in a thirty-six-year career. Its maiden voyage in 1876 included Frank A. Jordan of Foxcroft, until he deserted ship. By 1885, *John and Winthrop* sailed from San Francisco as a Shanghaier. It wintered at Herschel Island in 1894–95 and 1895–96. During its 1896 season, rumors of a gold strike in Yukon reached the winter quarters, and its first mass desertion took place the third week of January. This happened when boat-steerer Dan Sweeney led six other men away from ship. Other nearby captains thought more desertions might happen soon, so they guarded storehouses at night, allowing no one off ship after ten o'clock. Sweeney was recaptured, but others got away. He made good on his escape when he later stole a whaleboat and hid out as *John and Winthrop* sailed for home. He remained ashore in a sod hut and stirred up trouble all winter. *John and Winthrop* made one last whale voyage to the Arctic in 1912 but spent more time trading with Inuit. After a distinguished whaling career, it was laid up in Oakland, California, on October 28, 1912.[205]

Bath-built 1876 schooner *Sarah W. Hunt* made eight whaling voyages to the Atlantic and the Bering Strait, starting in 1883. Its 1886 voyage out of New Bedford included forty-four-year-old Barney Power of Portland, although it is recorded that he deserted at Callao in February 1886. That same voyage included Mainer James H. Smith. *Sarah W. Hunt* continued as a whaler until condemned and sold at Tenerife on October 21, 1896.[206]

Bath ship builders turned out at least five barks in 1877, and all eventually went whaling. Built by Goss & Sawyer, 348-ton *James Allen II* went straight into whaling and made a total of twelve voyages to north Atlantic and Arctic waters—the first three out of New Bedford and the remainder as a Shanghaier. It was reported lost off the Aleutian Islands on November 11, 1894. Its predecessor, *James Allen*, likely built in Bath, also had a distinguished whaling career. Its 1855 voyage included James Tyler of Portland, although he left at St. Catherines. Mainers Edmund Lewis of Medway, Ethan Shepley of Levant, Charles Van Wyck of Bangor and William Chandler and James R. Sloan of Eastport sailed aboard *James Allen* in 1859. Mainer Henry Cole and Richard Benke of Bangor were aboard for its 1872 voyage. Bath-built, 384-ton *Josephine*, also built by Goss

View of *Josephine* at wharf in New Bedford. *Maine Maritime Museum.*

& Sawyer, made twenty-three voyages to the Atlantic and Pacific Oceans, the north Pacific, the Sea of Okhotsk and Japanese waters. Between 1886 and 1891, it operated from San Francisco as a Shanghaier and was sold to Valparaiso owners in 1909, where its new Chilean owners also used it as a whaler. *Josephine* was one of the last of active whalers lost in 1919. Its career totals include 31,349 barrels of oil, 10,380 of which were sperm whale, and a staggering 180,350 pounds, or 90 tons, of baleen. *Josephine*'s binnacle is preserved at the New Bedford Whaling Museum.[207]

Another 1877 Goss & Sawyer bark that became a San Francisco Shanghaier was steam-converted 313-ton *Lucretia*. It made ten voyages to north Pacific waters but was not as productive as William Lewis's other steam whalers *Belvedere* and *North Star*. *Lucretia* proved a slow sailor and unlucky, as well. In a hurricane off Bermuda on its refitted maiden voyage, *Lucretia* barely survived and lost its topmasts, bowsprit, numerous spars and all boats but one. When it tried to round Cape Horn after repairs, adverse winds forced it to go the Indian Ocean route instead, and it arrived in San Francisco in November 1882 having missed the entire Arctic season. In June 1883, *Lucretia* harvested walruses, though the crew could only haul in

*Above*: Whaling bark *Josephine* discharging whale oil in New Bedford, circa 1909. *Penobscot Marine Museum.*

*Left*: *Josephine*'s binnacle at New Bedford Whaling Museum. *Photo by author.*

eighteen of forty or fifty that were shot. In 1888, it was one of first steam whalers to anchor on Herschel Island's east side. *Lucretia* smashed its way out of an ice trap in September, where Captain Edmund Kelley declared, "I'm not going to stay here and die like a rat in a hole." The ship made it home with fifteen tons of baleen. *Lucretia* was wrecked in the Chukchi Sea in September 1889.[208]

Bath's 188-ton bark *Mabel* was built by Brown & Hodgkins and sported a figurehead of a little girl named Mabel, the daughter of an owner. It had been carved by Civil War veteran and well-known Bath carver Colonel Charles Sampson. *Mabel* made five voyages to Indian, north Pacific and Arctic waters. On August 10, 1885, it was at ice edge near Wainwright Inlet between Point Belcher and Point Franklin, when a sudden gale blew up and parted a nearby ship's cable. The old whaler *George and Susan* dragged its other cable and ran afoul of *Mabel*—a collision that broke *George and Susan*'s main yard. *Mabel* then began to drag. Captain Healey of revenue cutter *Corwin* took the ship right into breakers for assistance and to offload crew, although he was unable to save it. Long-time whaler *George and Susan*'s 1827 voyage out of New Bedford included John Woods of Hollis, Maine. John H. Steward of Bath sailed in 1833, and Albert C. Field of Brunswick had gone whaling on it in 1857. Bath's 179-ton bark *Attleboro*, also by Brown & Hodgkins, made two voyages to Atlantic waters between 1877 and 1883. Its second voyage was under the command of three different captains at different times. *Attleboro* returned with 830 barrels of oil. It was lost off Mocha Island. In 1878, Bath schooner *Cameo* was built. It made seven whaling voyages to the Atlantic between 1908 and 1921. Its career totals include 10,991 barrels of sperm whale oil.[209]

*Mary and Helen*, Bath's new 420-ton steam bark, was finally launched from Goss, Sawyer & Packard's shipyard on July 29, 1879. Its figurehead of Mary Haskell of New Bedford and the image of her sister, Helen, on the stern was also carved by Colonel Sampson. The ship was expensive and looked it with a decorative figurehead and quarter board that matched an equally luxurious main cabin. On August 24, its steam plant drove it on a trial run at eight knots on Kennebec River. Owner William Lewis's goal was to pioneer steam power in Arctic whaling, and specially constructed *Mary and Helen* was his first attempt. It made one whaling voyage to the Arctic soon after launching, dramatically changing the efficiency of commercial whaling. Equipped with both sails and a steam engine, it operated off Alaskan shores without relying on changeable winds. It made nine knots on steam power alone. Such ships were now capable of directly following whales and remaining longer on hunting grounds. That voyage, begun under master Martin Van Buren Millard, who left ship at some point and was replaced by Maine native master Leander C. Owen, also included Mainers James Brown of Portland, Henry R. Farrin of Bath, Frank J. Smith of Bangor and Winfield Trott of Eastport. *Mary and Helen* returned to San Francisco in October 1880 with John Wykoff of Revenue Cutter service, who reported that cutter *Corwin* had been unable to reach Herald Island due to heavy ice. It also returned with 265 barrels of sperm whale oil worth $28 a barrel, 2,350 barrels of whale oil at $16 per barrel and forty-five thousand pounds of baleen worth $2 per pound. *Mary and Helen* was eventually sold to the government for $100,000 and renamed USS *Rodgers*. It participated in a search for survivors of the failed Arctic *Jeannette* Expedition. It was destroyed by fire on November 30, 1881.[210]

Another vessel built in 1879 that found its way into whaling was Kennebunk schooner *Wanderer*. It made only one voyage to the south Atlantic in 1880. Departing New London in August under master Walter Scott Chesebro, the crew included twenty-two-year-old W.H. Collins of Portland. *Wanderer* was lost in the Straits of Magellan on October 17, 1881.[211]

Whaling steam-bark *Mary and Helen*. *Maine Maritime Museum.*

In 1880, 508-ton Bath steam-bark *Belvedere* went right into whaling. As elegant as *Mary and Helen*, this whaler had clipper-like lines, an eagle figurehead and a rounded stern. *Belvedere* made four voyages to the Arctic from 1880 to 1883, first with Maine captain Leander Owen. It was a highly successful voyage, as Owen hauled in six hundred walruses and fourteen bowheads, which produced 1,800 barrels of oil, seventeen tons of baleen and 2,500 pounds of walrus ivory. *Belvedere* also returned with news of the loss of USS *Rodgers*, formerly *Mary and Helen*. That 1880 voyage also included Mainers Herbert C. Shea and O.W. Stacy, both of Bath. From 1886 to 1890, *Belvedere* made four more Arctic voyages, this time as a Shanghaier. Its haul during those four voyages included 2,740 barrels of oil and almost twenty-nine tons of baleen. It wintered at Herschel Island in 1895–96 under captain Joseph A. Whiteside, and another early freeze trapped whalers north of Cape Barrow. Aboard *Belvedere*, one of the mates, (later captain) George F. Tilton, volunteered to get word out of their status, inform *Belvedere*'s owners and send supplies for next season's whaling operations. With a native husband-and-wife pair to help guide, Tilton left with a dog team and fifteen days' worth of rations. He crossed Kotzebue Sound to the mouth of Buckland River on to Norton Sound. He was constantly forced to test weak ice with an axe that he carried until it slipped from his numbed hands and disappeared into the ocean. Tilton got around broken ice at Cape Lisburne but froze his fingers and toes in a blizzard. He found an old dory at Katmai, used his underwear to make it seaworthy and then rowed thirty-five miles across Shelikoff Straits and landed safely just as a blizzard hit. When he arrived at Seattle, ship owners did not believe his harrowing tale and thought him a deserter from the fleet, but word of the whalers' plight got out and help was sent.[212]

In 1897–98, *Belvedere* was once again trapped by ice and wintered at Peard Bay, Alaska. There, it rescued crews of *Jesse H. Freeman* and *Orca*. *Belvedere* eventually escaped with extensive damage to a sprung rudder. It reached Port Clarence on August 4, where it got coal and then returned to Point Barrow to drop off supplies for other trapped ships. *Belvedere* was again trapped but survived the winter safe behind heavy ground ice at Sea Horse Islands. It wintered at Baillie Islands in 1905–6 under captain S.F. Cottle. By 1912, even Captain Cottle accepted the inevitable about the declining industry, stating, "It is no use to pay out three or four dollars for every pound of bone you catch and then receive $1.50 or $2.00 a pound." *Belvedere*'s career in north Pacific and Arctic waters lasted until it was crushed by ice off Siberia in 1919.[213]

Steam-bark whaler *Belvedere* tied at wharf. *Maine Maritime Museum.*

In 1881, Bath steam bark, 508-ton *Mary and Helen II* was built and made six whaling voyages to north Pacific, Arctic and Sea of Okhotsk waters. Its 1882 voyage included George C. Hamm of Portland and Elias W. Hogan of Bath, under master George Fowler Smith. They returned with 300 barrels of whale oil and 4,500 pounds of baleen. It was majorly damaged by fire at San Francisco after that 1882 voyage but was rebuilt. *Mary and Helen II* continued whaling until 1886. The last voyage as *Mary and Helen II* netted almost 6 tons of baleen. It was renamed *Beluga* by 1887 and made fourteen more whaling voyages to the north Pacific. As *Beluga*, it made voyages as a Shanghaier out of San Francisco. The first three were under master Joseph Franklin Brooks. *Beluga* sailed to the north Pacific and returned with 2,675 barrels of oil and over twenty-five tons of baleen. In 1888, it proceeded eastward in company with revenue cutter *Thetis* to Herschel Island. That season, *Beluga* rescued the crew of *Young Phoenix* after it had been crushed in ice. In 1890, *Beluga* reported taking a whale with an iron in it from whaleship *Montezuma*, which had last been whaling in 1853. *Beluga* wintered at Herschel Island in 1894–95 and 1895–96, at Langton Bay in 1897–98 and 1904–5 and at Baillie Islands in 1898–1900 and 1905–6, when it was trapped by ice after intending to winter at Herschel Island. That first winter at Herschel Island included master Albert C. Sherman's wife and young son Bertie. They hosted the season's first party—a deck housewarming. To avoid boredom during long winters, the crew put on elaborate theatricals and constructed a stage, complete with footlights on the poop deck—the largest enclosed space available on the island. They even printed programs and handed them out to audience members and had separate performances for crews and natives, as well as officers and engineers. Space was used for Sunday services and a baptism, when Mrs. Sherman gave birth to a baby girl named Helen Herschel. *Beluga*'s 1899–1900 voyage was considered to be the last good catch of an American whale fishery, with fifty-six tons of baleen.[214]

In 1900, *Beluga* was used to launch specially constructed drift casks off Banks Land to determine icepack and current movements in the Arctic Ocean. It wintered at Baillie Island under master Hartson Hartlett Bodfish, who bought two musk ox calves that were harnessed to a sled. One was killed by dogs, but Bodfish sold the other, named Olive Jones, to Bronx Zoo for $1,600. That season, *Beluga* returned with 425 barrels of whale oil and twenty-two thousand pounds of baleen. In 1907, *Beluga* went to the Pacific under master George W. Porter. It was reported as caught in ice north of Alaska. Its last voyage as a whaler in 1908 brought back five thousand pounds of baleen. In July 1917, the hard-working steam bark whaler was sunk by a German U-boat during World War I.[215]

In 1881, Bath's 489-ton steam bark *North Star* was built and launched by Goss, Sawyer & Packard. Its maiden voyage started under master James S. Carter, but he left ship at some point and was replaced by Mainer Leander C. Owen. On July 8, 1882, *North Star* was crushed by ice three miles off Cape Smyth, Alaska, near Point Barrow. It had been crushed by force of ice "so great that the cracking of her timbers could be heard on shore." The owner, a Captain Lewis, was financially devastated and ended up selling his latest whaler, *Mary and Helen II*, to San Francisco competitors to recoup his losses.[216]

In 1883, Bath built a 512-ton steam bark named *Thrasher*, as ordered by San Francisco interests who had formed Pacific Steam Whaling Company. It made twenty-eight whaling voyages out of San Francisco to the north Pacific as a Shanghaier. In 1888, it was the first steam whaler to drop anchor on Herschel Island's east side. *Thrasher* wintered there again in 1894–95, with Captain C.E. Weeks and his wife aboard. It was a difficult winter, and Captain Weeks was killed after falling into the ship's hold. James A. Tilton replaced him as master. *Thrasher* took part in rescuing Charles D. Brower and survivors from the *Navarch* disaster in 1897. In September 1899,

*Thrasher* participated in a scientific venture that involved putting specially constructed casks adrift in the Arctic Ocean to determine currents and icepack movement. At a specified location west-northwest of Point Barrow, it set one cask adrift. Later, *Thrasher* was trapped by ice and had to winter once more at Herschel Island in 1905–6. It was reported as caught in ice off northern Alaska in November 1907. In 1908, it was sold to Seattle owners and renamed *Kamchatka.* Put into Alaskan trade, it caught fire and was lost in 1921 in the Aleutian Islands.[217]

Another Bath steam bark built in 1883 was 516-ton *Jesse H. Freeman.* Launched March 29, it was designed to carry ten thousand bunches of bananas from the West Indies to Boston for United Fruit Company in ten to twelve days. Once it began whaling, it made eight voyages out of San Francisco to the north Pacific as a Shanghaier. In 1888, *Jesse H. Freeman* anchored at Herschel Island's east end with alarming news that there was a shoal fifteen miles offshore. It wintered there again in 1894–95 and 1895–96. That first winter, it was under master William P.S. Porter, who brought along his twenty-nine-year-old wife, Sophie, and daughter, Dorothy. In 1896–97, *Jesse H. Freeman* was trapped in ice on its way to Herschel Island and wintered instead near King Point. Captain Porter and half of its crew left the ship in ice and wintered on Herschel Island. Evening entertainment included violin and piano recitals and a usual round of dinners with other masters. *Jesse H. Freeman*, *Belvedere* and six other whalers were caught in ice in 1897. Most were lost, including *Jesse H. Freeman*, when it was crushed off Sea Horse Islands, Alaska, on September 22, 1897—the same season of *Navarch*'s disaster. *Belvedere* rescued its crew. One source states that natives set fire to *Jesse H. Freeman* and burned it. It was a total loss, and captain and crew wintered with men of *Belvedere.*[218] The captain wrote, "All the yards were lit up with the fire, and once in a while a 100 gallon drum of coal oil would explode and the flames would shoot up. We could hear the cracking of the bombs that were to have been used for killing the whales. It sounded like a New Year celebration in Chinatown."[219]

In 1883, Boothbay schooner *Bertha D. Nickerson* was built. It made six voyages out of New Bedford to the Atlantic from 1906 to 1915. At least one of those voyages included H.G. Spencer of Belgrade, Maine. *Bertha D. Nickerson*'s best hauls of sperm whale oil were its first and last seasons. In 1908, it brought back 2,130 barrels and 1,350 barrels in 1915. Its six-season totals included 4,880 barrels of sperm whale oil.[220]

Bath steam barkentine *Morning Star* was built in 1884 and made twenty whaling voyages out of San Francisco to the north Pacific starting in 1903.

It also helped supply mission outposts in the Pacific Islands. Mainers Frank W. Proctor of Somerset and James L.G. Wright of Portland shipped aboard it. In 1904, its name was changed to *Herman*, and it continued whaling out of San Francisco. In 1907, *Herman* was reported as caught in ice off Alaska's north coast. It continued whaling until 1924, at which point it was sold to foreign owners. It disappeared in 1923 in Mexican waters as motor ship *Chapultepec*. Kennebunk's 109-ton schooner *Mattie T. Dyer* was built in 1884. By 1890, it operated out of San Francisco as a Shanghaier but went after mostly seals and otter under master C.E. Ockler. There is no record of it in NMDL's database.[221]

In 1886, 325-ton Bath bark *William Baylies* was built by New England Shipbuilding Company and went straight into whaling. It was rated A-1 for fifteen years and was designed first for sailing only. Its keel was made of maple, oak and birch, and its stern, sternpost and stanchions were made of oak. It had galvanized fastenings throughout the hull and keel bottom. On one voyage, it tore off fifty feet of metal on the keel but sprang no leak. *William*

Whaling ship *Morning Star*. *Maine Maritime Museum.*

*Baylies* made twenty-two voyages to north Pacific and Arctic waters—the first five out of San Francisco as a Shanghaier. Underpowered, it was converted to steam in 1894 and then wintered in the Arctic once, at Herschel Island in 1894–95 under captain John McInnis. *William Baylies* returned in late 1897 with news of eight other whalers imprisoned in ice east of Point Barrow. In 1907, it was reported as caught in ice, but is successfully escaped. In 1908, *William Baylies* was fast to drift ice in company with whaler *Bowhead* when the floe collided with land ice some distance from shore. Caught between two ice floes, it was "crushed like an eggshell" in Anadir Sea on May 16, 1908.[222]

*Jane Gray* was a 113-ton Bath schooner built in 1887. It made six yearly whaling voyages to the north Pacific out of San Francisco as a Shanghaier. In 1888, during a storm, it was almost rammed by drifting bark *Eliza*. To avoid collision, *Jane Gray* slipped its cable and ran off downwind, nearly smashing into a third vessel, *Bounding Billow*. *Eliza* was not as fortunate and stove in *Bounding Billow*'s bulwarks. *Jane Gray* continued out of control in the wind and crashed into an iceberg, which stove a great hole in its starboard side, into which water poured in a great stream. It could not be stemmed with canvas, and the captain maneuvered the vessel as best he could to nearby ship *Andrew Hicks*. The crew abandoned *Jane Gray*, and she sank. They safely reached another vessel in whaleboats.[223]

In 1888, Bath's 463-ton steam bark *William Lewis* was built by Kelley & Spear, although another source says it was built in 1891. The hull was reinforced with five sets of diagonal braces. *William Lewis* made six whaling voyages to the north Pacific out of San Francisco as a Shanghaier. Its 1888 voyage included Willis E. McCagney of Bath, who was discharged at some point along the way. The crew started under master Cyrus Manter, but Albert C. Sherman replaced him. Sherman returned with 175 barrels of whale oil and five thousand pounds of baleen. *William Lewis* ran aground at Point Barrow on October 3, 1891—the result of a piloting error.[224]

Bath's 494-ton, 142-foot steam bark *Navarch* was built in 1891 or 1892 for New Bedford whaleman William Lewis, a pioneer of Arctic steam whaling. It had a 320-horsepower auxiliary engine and sported a Hyde steering apparatus that protected the propeller when backing into ice. *Navarch* made three voyages to the north Pacific between 1892 and 1897. Its first, in 1892, included Mainer E.L. Benson, who signed on as seaman. That season, *Navarch* was reported to have returned leaking. It wintered at Herschel Island in 1894–95 and 1895–96, the first winter with Captain John A. Cook's wife aboard. *Navarch* was the first whaler to reach Icy Cape in the 1897 season. On July 27, Captain Joseph A. Whiteside ordered the anchor set and welcomed aboard Charles D. Brower, who had come 150 miles along the coast from Point Barrow to rendezvous with arriving whaleships. In his memoirs, Brower recorded his first meeting with captain, whom he called Whitesides: "the man whose extraordinary behavior was to go down through the years as an ugly blot on my memories of the sea." According to Brower, Whiteside decided to move *Navarch*'s anchorage to the north side of Icy Cape when he learned whaleship *Karluk* had also arrived and anchored nearby. This was ostensibly to keep Whiteside's wife happy. She seemed to harbor a long-standing grudge against the wife of *Karluk*'s captain. Separated, they would have no chance to "gam" or socialize, but the decision put *Navarch* in a precarious position on north side of Icy Cape, especially with a west wind, was dangerous.[225]

Whiteside had captained other Maine-built whalers, most notably *Lucretia* and *Belvedere*. He joined *Navarch* in San Francisco in March 1897 as replacement and may have had some premonitions about his upcoming voyage before he left Massachusetts. According to New Bedford acquaintances, Whiteside confided that he thought it might be time for

*Navarch. Maine Maritime Museum.*

his years of good luck to turn.[226] He was quite prescient. When Whiteside repositioned *Navarch* to mollify his wife, he put the ship into a more exposed position. Contrary winds came up and packed ice cakes around *Navarch*. Brower had stayed aboard *Navarch* for a return trip north to his home at Point Barrow but now expressed caution to Whiteside. "I had seen the insidious, creeping action of the ice too often not to call Captain Whiteside's attention to the danger….But instead of steaming immediately south of the shoals again, he seemed to think it safe to tie up to a large floe, apparently grounded. So we stayed there, while drifting ice packed in solid on all sides."[227]

Grounded floe broke loose and moved rapidly north, taking *Navarch* with it. In his memoirs, Brower suggested that Whiteside froze that moment by not immediately trying to break free. In hours, it was too late, and *Navarch* drifted northeast "as a fly glued to a sheet of floating flypaper." It drifted thirty or forty miles, and *Navarch* was then caught in a three-mile-per-hour current, which swept it quickly north. By August 2, it was off Sea Horse Islands, hopelessly trapped. Whiteside decided to abandon ship and haul whaleboats with supplies across ice to open water. Brower suggested building false keels on the bottoms of whaleboats to withstand the rough ice that they were dragged over, but Whiteside declined to heed Brower's advice, as he did not want to add more weight to haul. It was an unfortunate decision. Dragged across rough ice, bottom boards were soon torn out on two of three whaleboats, and they had not brought any canvas to repair them.[228]

Brower wrote that Whiteside seemed to lose his head at this point. When offered advice, he often took the opposite tack. When the remaining two boats were smashed beyond use, Whiteside broke down and cried. To Brower's disgust, Whiteside then produced a bottle and began drinking. He offered men gold if they would get him ashore. They decided to return to *Navarch*, but Whiteside ordered everyone to fend for themselves and set out with some stronger crewmen. Brower was shocked that Whiteside abandoned his weaker men and Whiteside's own wife, whom Brower now helped along. After a desperate attempt to get back to *Navarch*, they realized they would have to abandon it once more and make their way to land before they drifted too far north past Point Barrow, never to return. They built a small hickory-framed boat that was light enough for two to haul over rough ice yet sturdy enough to ferry up to ten at a time across open leads. Whiteside ordered Brower to start with some strong men to open trail. He would come along behind with a boat and supplies. When they reached a lead of open water, Brower waited for the boat and supplies. But Whiteside was nowhere to be seen—he had retreated back to *Navarch* with a few men. Now, exhausted men with Brower had no boat and no food.[229]

*Navarch*'s steward, W.W. Whiting, became despondent and shot himself. According to Brower, "this was bad for everyone." They left him where he lay and hurried south as fast as they could manage. On August 12, two gave up and asked to be left behind. They were never seen again. *Navarch*'s chief engineer, named Sands, then became unstable and had to be restrained. He died the following day, and two firemen asked to be left behind. Some remaining men became snow blind, and their constantly wet clothes began to deteriorate. A Portuguese crew member was next to ask to be left behind. Brower was haunted by those horrible fifth and sixth days of their march: "It was heart-rending to see those fellows drop back one by one, hear their cries fade out in the distance. Hardest for me to bear was the fact that they always called, 'Charlie! Charlie! Don't leave me, Charlie!' Their voices rang through my head for years."[230]

They managed to get on an ice floe and drifted. At one point, a vessel passed near enough for them to hope for rescue, but they were not spotted. They had been without much food for eleven days, and many men broke down when they were finally spotted by whaleship *Thrasher* and taken to Point Barrow. Thirteen of twenty-nine men perished. Whiteside managed to get himself, his wife, the small boat and remaining men with him to shore at Cooper's Island, where they were found by revenue cutter *Bear*. Brower wrote that they were all in excellent condition. When asked why he had turned back from Brower and most of his crew, Whiteside said nothing and just turned away. Many blamed him for the disaster, and he never commanded another whaling ship. Joseph A. Whiteside died in three years.[231]

Four crew had remained aboard *Navarch*, and at least one of them was convinced that they could claim salvage rights to the abandoned vessel. They were eventually rescued in mid-October, by whaleships *Newport* and *Fearless*, which had been forced to winter in ice and were spotted when *Navarch* drifted twelve miles off Point Tangent. As they were low on supplies, other crews approached *Navarch* hoping to make use of its stores. Men aboard *Navarch* said they could take some provisions but only with a written acknowledgement that payment would be made to rightful owners of the stores. This was refused, and crews loaded sleds with provisions anyway but lost them on their return to their respective ships. Masters of *Newport* and *Fearless* determined to bring over the rest of *Navarch*'s stores without written acknowledgement or requested receipt. Assistance was given by some nearby native inhabitants, who were also promised some of the abandoned whaler's equipment in return for help. The master of *Newport* was later sued for these trespasses and ended up having to pay for stores and equipment that were ordered removed from *Navarch*.[232]

*Navarch* remained frozen in ice but was eventually driven ashore near Point Barrow. Hoping to get its supply of coal, men boarded the ship, but before it could be offloaded, *Navarch* was accidentally set on fire. It burned down to the keel. One source states that the fire was started by two workers who got tired while unloading coal. The dramatic disaster was considered one of Alaska's worst shipping losses of all time and was commemorated in a scrimshaw by artist Salman Rashidi in the 1990s.[233]

In 1899, Boothbay built the schooner *Valkyria*. It made twelve whaling voyages out of Portland and a final one out of New Bedford to Atlantic waters spanning the years 1909 to 1920. Its career totals include 5,724 barrels of sperm whale oil. The 139-ton brig *Viola* was built in Essex, Massachusetts, in 1910. It made three separate whaling voyages also out of Portland to Atlantic waters. Most were under master John Atkins Cook or master Edwin J. Reed. Under these men, *Viola* returned with 7,130 barrels of sperm whale oil. In 1918, it was reported missing.[234]

*Adze and hammer and anvil stroke*
*Echo not from the shore;*
*The wharves are old and broken and gray*
*And the whaleships come no more.*[235]

# 7.

# FINIS

The American whaling industry made its greatest profits along the northwest coast of America from 1845 to 1855. At that time, it was controlled by Russia. It is argued that even with great profits, such activity had little influence on U.S. federal legislation or national policy. In fact, between 1835 and 1845, not one single piece of federal legislation offered or discussed in Congress dealt solely with whaling off America's northwest coast. But for Russia, such extensive American activity led the country to consider eventually ceding or selling its territory. From 1835 until Alaska's purchase in 1867, nearly 60 percent of all oil was secured from America's northwest coast. This increasingly came from walruses and bowhead whales. Pacific walruses became an important part of the catch, mostly after 1849, when whaling vessels started passing north of the Aleutians and Commander Islands. But it was not until the 1860s that a relatively steady market for animal oils and decreasing numbers of available bowhead whales brought about deliberate walrus hunting. This intensive harvesting of walruses lasted into the early 1880s, when depletion of stock and declining prices for oil made hunts unprofitable. Those seventeen years saw 90 percent of the walrus harvest.[236]

American whaling began to seriously decline by the 1860s. The ravages of the Civil War on American shipping resulted in the loss of twenty-four whaling ships captured and burned by Confederates. Other factors included a developing petroleum industry, declining price of whale oil, growing scarcity of whales and rising industry expenses. The Crimean

War in Europe, economic depression of 1857 and a growing availability of cheaper substitutes for whale oil as illuminants, like kerosene, also strained the industry. A series of major freezes and subsequent losses of whale ships in Arctic disasters in the 1870s and 1890s decimated the Pacific whaling fleet. From a high of six hundred whale ships in 1855 to a low of forty by 1879, whaling was in serious decline. The *New York Times* noted only eighteen vessels fished north Pacific and Arctic waters, and they put in at San Francisco rather than the temporary halt they formerly made at Hawaii. While the *New York Times* extolled the efficiency of whalers, taking three of four whales encountered, the article predicted tougher times ahead.[237]

An article in November 1883 recognized a somewhat reinvigorated industry, but within twelve years, it lamented the passing of an era by noting America's entire whaling fleet had dwindled to a paltry seventy-three vessels. The final and likely fatal blows were events of 1897 through 1898, when several whaleships, including Maine-built *Navarch*, were either lost or trapped in ice off northern Alaska. This necessitated a unique and epic rescue attempt in which three officers of the U.S. Revenue Cutter service herded reindeer eight hundred miles to feed the stranded men at Point Barrow. By 1907, prices for baleen had dropped from seven dollars a pound to less than half a dollar. Sources state that by 1910, the market had entirely disappeared, although some whaling stations along Alaska's southeast coast managed to stay in business until the 1930s. With the loss of Maine-built whaleship *Wanderer* in 1924, on its first day out on its last whaling voyage, the *New York Times* called it the end of a great chapter, even though that very newspaper had, a month earlier, written of new commercial whaling industry possibilities using airplanes. *Wanderer* was the last three-masted whaler to sail out of New Bedford.[238]

This newer whaling industry the *New York Times* spoke of also had Maine connections. In 1869, Charles Bliss built 1,065-ton bark *Jennie S. Barker* in Freeport. It participated in the American cotton trade, survived numerous transatlantic crossings and difficult Cape Horn passages, suffered a near-fatal and far-reaching mid-channel collision, toiled for years in the Baltic timber trade and witnessed a South American naval war. In 1904, after thirty-five lively years as a merchant, it became available to Carl Anton Larsen as it sat idle in Sandefjord, Norway. Larsen, interested in launching commercial whaling in the Antarctic, bought and renamed it *Louise*. It made a one-way voyage to Grytviken Harbor, South Georgia Island, and was used for housing, general stores, coal storage and then a mooring hulk. Abandoned, it was towed out of the way of the thriving onshore whaling station and left

*Louise* near Grytviken cemetery, South Georgia Island. *Photo by John Splettstoesser, courtesy of Antarctican Society.*

to rot next to Grytviken's small cemetery, where polar explorer Sir Ernest Shackleton is buried. Inadvertently burned to the waterline in the early 1980s, its remains still sit near the now-defunct whaling station, one of the last surviving examples of a Maine-built "down-easter."[239]

One other Maine-built vessel to involve itself in this more modern phase of the whaling and sealing industry was a fleet auxiliary tug launched into World War II waters of Rockland Harbor on September 2, 1944. Built by master builder Eliot Gamage at Snow Shipyard, it was eventually named *ATA-215*. It displaced 1,500 tons and was 194 feet long. Its propulsion was a single propeller powered by Busch Sulzer 539 diesel-electric engine with two 60-kilowatt, 120-volt DC Ship's service generators, which gave it a top speed of 12.1 knots. Its cruising range was limited to the 935 barrels (or 39,270 gallons) of diesel fuel it could carry. It saw wartime service in the Pacific, and then polar explorer Finn Ronne used it for his expedition to the Antarctic Peninsula in 1947. Since the federal government did not allow the U.S. Navy to lend ships to individuals or organizations, even for polar exploration, Congress had to get involved. The Navy "loaned" *ATA-215* to Ronne, and President Harry Truman signed the bill into law. By early 1947, it was in Beaumont, Texas, and rechristened *Port of Beaumont*. Jennie

*Louise* in 2012. *Photo by John Splettstoesser, courtesy of Antarctican Society.*

Darlington, wife of one of the expedition's pilots, recorded that the ship wore the resigned air of a beast of burden on whose back tourists are being given a ride. Expedition member Robert H. Dodson called it a fine ship—seaworthy but not speedy.[240] He was in charge of the expedition's dogs and recalled that while crossing the tropics, flying fish came through the hawser holes, much to the delight of the dogs: "Ship was pretty low. We didn't have lots of freeboard, low freeboard and these big hawser holes and the flying fish would come right through and the dogs would eat them."[241]

*Port of Beaumont* became the first diesel-driven ship intentionally frozen into Antarctic ice for the winter. As unloading began and the shore base was being built, most expedition members remained aboard until early May. Then *Port of Beaumont* was used as a storehouse. Dodson remembered that they were constantly going back and forth to it. In February 1948, they needed help extricating themselves from ice, so two U.S. Navy icebreakers in the area arrived to help.[242]

Dodson recalled, "It was an ATA—whatever that meant—sea-going tug. It was all wood. That was a big factor. Flexible to the ice pressure. And it was very sea worthy. Diesel engine. 1,200 tons, I think. 50-foot beam. Does that seem too small? Less than that….And it had spent the war hauling, oh,

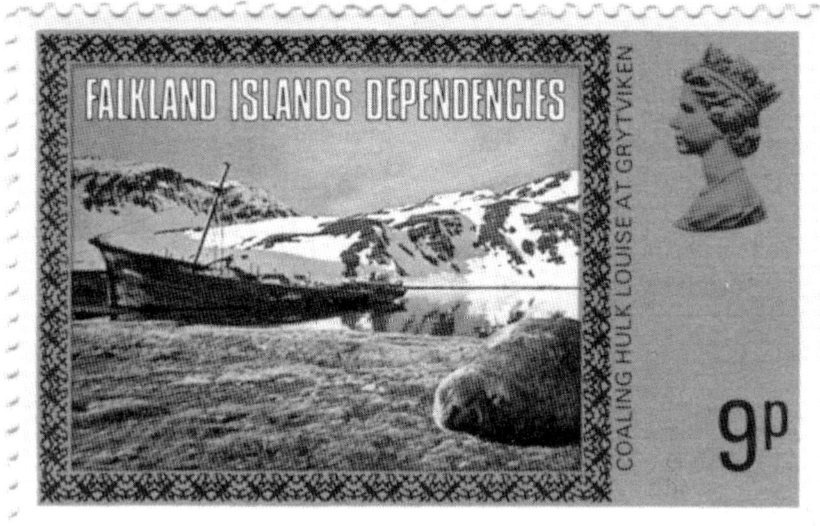


*Louise* (previously *Jennie S. Barker*) commemorated on Falkland Islands Dependencies postage stamp. *Author's collection.*

*ATA-215* (later *Port of Beaumont*, later *Arctic Sealer*) after 1944 launching. *19-N-76424, National Archives and Records Administration.*

barges it was across the Pacific. A lot of low gear power. It wasn't fast, but it had a lot of power. Built in Rockland, Maine, where they know how to build the old schooners."[243]

After Ronne's return, *Port of Beaumont* was sold to Canadian owners Shaw Steamship Company of Halifax, Nova Scotia. Put into sealing, it was renamed *Arctic Sealer*.[244]

After the 1957 season catch, *Arctic Sealer* was put into service by Canadian authorities as part of Operation Look See, a joint United States–Canadian Air Defense Command mission that brought aircraft, pilots, crewmen, civilian engineers and contractors to Greenland to complete final surveys of Distant Early Warning (DEW) radar sites. In 1958, it was chartered by Canadian Hydrographic Service (CHS) for upcoming surveying seasons. In 1961, it went under charter to the Bedford Institute of Oceanography as *MV Arctic Sealer* and surveyed Newfoundland harbors, such as Hawkes Bay, Nain and Lake Harbour on Baffin Island.[245]

*Arctic Sealer*'s final days, however, are sketchy. It sank in April 1963. It is not clear if *Arctic Sealer* was still being chartered by CHS at that time, though a news item in the London *Times* reported *Arctic Sealer* as a British ship. Dodson thinks *Arctic Sealer* ended its days in the Arctic in the late winter or spring in

*Port of Beaumont* frozen in for winter, Stonington Island, Antarctic Peninsula. *Photo by Robert H.T. Dodson, courtesy of Antarctican Society.*

the 1960s and suggests "her rotting old wood a cause no doubt," as she sank in Canada Bay, Newfoundland.[246]

By the end of World War I, and into the 1920s, American whalers had hunted, sealed, wintered, sailed and wrecked in just about every corner of the globe. But the era had ended. It is written that the huge amount of oil obtained provided the fuel for the lighting and lubrication for gears of America's Industrial Revolution. At its height in the mid-1840s, American whaling had ranked ninth in overall value to the economy.[247] By the mid- to late twentieth century, however, American involvement had dwindled. Long gone were those glorious days of American whaling when the towns, men and ships of Maine had been involved in practically every step along the way.

*We're homeward bound to New Bedford town;*
*Good-by, fare you well; good-by, fare you well;*
*When we get there we will walk around;*
*Hurrah, my boys, we're homeward bound.*

*And now our ship is full, my boys;*
*Good-by, fare you well; good-by, fare you well;*
*We'll think of home and all its joys;*
*Hurrah, my boys, we're homeward bound.*

*It's when you see those New Bedford girls;*
*Good-by, fare you well; good-by, fare you well;*
*With their bright blue eyes and flowing curls;*
*Hurrah, my boys we're homeward bound.*

*When we're paid off, we'll have a good time;*
*Good-by, fare you well; good-by, fare you well;*
*The sparking of girls and the drinking of wine;*
*Hurrah, my boys, we're homeward bound.*

*We'll spend our money free when we're on shore;*
*Good-by, fare you well; good-by, fare you well;*
*And when it's all gone, we'll to sea for more;*
*Hurrah, my boys, we're homeward bound.*[248]

# NOTES

## *Preface*

1. Leland, *Algonquin Legends*, 35, 75; Stearns, *Story of the New England Whalers*, 3, 6, 21–22, 27–28.
2. "What About Whaling?" *Penobscot Marine Museum*; Hallowell and Wilson, "Whaling."
3. Stearns, *Story of the New England Whalers*, 3; "What About Whaling?"; Hallowell and Wilson, "Whaling."
4. Stearns, *Story of the New England Whalers*, 111–12; Bunting, *Live Yankees*, 9; Martin, *Whalemen and Whaleships*, 7.
5. Stearns, *Story of the New England Whalers*, 64, 111–12; Bunting, *Live Yankees*, 9; Martin, *Whalemen and Whaleships*, 7.
6. "Captain Alexander Coffin, Letter to Samuel Adams—July 8, 1785," in Stearns, *Story of the New England Whalers*, 109; Bunting, *Live Yankees*, 9; Martin, *Whalemen and Whaleships of Maine*, 7; "History of Captain John O. Morse and His House"; National Maritime Digital Library (NMDL) vessel: *Rhine*.
7. *Republican Journal*, February 2, 1888; Megan Pinette, personal communication.
8. "Blubber Knife," Maine Memory Network; "Sperm Oil Lamp, Circa 1790," Maine Memory Network.
9. NMDL vessel: *John Dawson*; "Sylvanus Baker," Whaling History; "Whaling History," New Bedford Whaling Museum (hereafter cited as NBWM).

10. Thaler, "Menhaden of History"; Webb, "Menhaden Whalemen," 277; Goode, *Fisheries and Fishery Industries*, "'Bony-Fish,'" 26.
11. Glover M. Allen, "Whalebone Whales of New England," 228; Luther Maddocks, *Looking Backward*, 42; Larkin, "Luther Maddocks' Fabulous Whale," 424; Webb, "Menhaden Whalemen," 279, 281–85; *Bath Daily Times*, May 11, 1880.
12. Stearns, *Story of the New England Whalers*, 111–12; Moment, "Business of Whaling," 269.

## Introduction

13. Davis, Gallman and Hutchins, "Technology, Productivity, and Profits," 147; Scammon, *Marine Mammals of the North-western Coast*, 211; Bockstoce, *Whales, Ice & Men*, 94.
14. "Whaling History," NBWM; Mrantz, *Whaling Days in Old Hawaii*, 9–10, 35; Walkerman, "Captain's Nightmare," 4–5.
15. Moment, "Business of Whaling," 268; "Hairpin, Circa 1840," Maine Memory Network; Davis, Gallman and Hutchins, "Technology, Productivity, and Profits," 147; Starbuck, *History of the American Whale.*
16. Martin, *Whalemen and Whaleships of Maine*, 17.

## Chapter 1

17. Field, *God's Pocket*, 144–45, 147–51, 161–63; Stanley, *List of Vessels Built*, 8; Good, personal communication; Proctor, *Fishermen's Own Book.*
18. Martin, "Voyages of Maine Whaleships," 46; Dorsett, "Maine Whalers," 14.
19. "Portsmouth Fails at Whaling"; NMDL vessel: *Triton.*
20. Dorsett, "Maine Whalers," 39; NMDL vessels: *Science, Canton*; "National Archives Project"; Norton, *Journal of a Voyage*; Martin, *Whalemen and Whaleships*, 17, 23; Starbuck, *History of the American Whale*, 314–15.
21. NMDL vessel: *Science*; Gifford, *Remarks on Board Ship ADDISON*; *Eastern Argus*, September 5, 1838; Starbuck, *History of the American Whale*, 352–53; Densmore, *Halo*, 145.
22. NMDL vessel: *Wiscassett*; Lipfert, "Maine Whalers"; Dorsett, "Maine Whalers," 15; Rice, *Shipping Days of Old Boothbay*, 165, 167; Fannie Chase, *Wiscasset in Pownalborough*, 417; Martin, "Voyages of Maine

Whaleships," 47; Martin, *Whalemen and Whaleships*, 26–27; *Journal of a Voige to the Pacific Ocean*.

23. Martin, *Whalemen and Whaleships*, 29; "Case Containing Whale Tooth," Andrew Carnegie Archives.
24. Starbuck, *History of the American Whale*, 314–15; Martin, "Voyages of Maine Whaleships," 48; Martin, *Whalemen and Whaleships*, 29; Fannie Chase, *Wiscasset in Pownalborough*, 418; Hawes, *Whaling*, 152; Humiston, *Blue Water Men*, 33; Dorsett, "Maine Whalers," 38.
25. Rice, *Shipping Days of Old Boothbay*, 146, 166; Martin, *Whalemen and Whaleships*, 30–31; Starbuck, *History of the American Whale*, 352–53; Martin, "Voyages of Maine Whaleships," 49; Rowe, *Maritime History of Maine*, 270; Hawes, *Whaling*, 152–53; Day, "Gil Whitman's Anvil Art," 49–50; Cutler, *Queens of the Western Ocean*, 515; Fannie Chase, *Wiscasset in Pownalborough*, 419; *Lincoln County News*, January 29, 1920; Biscoe, *No Pluckier Set of Men*, 63.
26. Martin, "Spirit Voices," 43.
27. Densmore, *Halo*, 82–83; Martin, *Whalemen and Whaleships*, 21–22.
28. NMDL vessels: *Peru, Henry, Massasoit*; Densmore, *Halo*, 139, 142; Martin, *Whalemen and Whaleships*, 33; Martin "Spirit Voices," 44.
29. Martin "Spirit Voices," 45; Densmore, *Halo*, 151, 159, 163, 164–65, 168, 179; Martin, *Whalemen and Whaleships*, 34–35.
30. Martin, "Spirit Voices," 65.
31. NMDL vessel: *Warwick*; Martin, *Whalemen and Whaleships*, 37.
32. *Eastern Argus*, June 2, 1841; Martin, *Whalemen and Whaleships*, 37–38.
33. NMDL vessel: *Warwick*; Starbuck, *History of the American Whale*, 386–87; Martin, *Whalemen and Whaleships*, 38; Martin, "Voyages of Maine Whaleships," 75.
34. NMDL vessels: *Beulah, Elbe*; Neil, *Journal Kept Aboard the Ship ELBE*; Martin, *Whalemen and Whaleships*, 39.
35. NMDL vessels: *Washington Freeman, Glacier, Georgia*; "Whaling History," NBWM; Lipfert, "Maine Whalers"; Kinney, *Vessels of Way Down East*, 135.

## *Chapter 2*

36. A. Bowdoin Van Riper, personal communication, June 2016.
37. Laing, *Seafaring America*; Thwing, *Crooked and Narrow Streets*, 182.
38. NMDL Voyage ID 100198 and 100201; "List of Vessels," 602–3; *Lloyd's Register of Shipping*, 23; NMDL. 1797–1804 Whaling Vessels; "French Spoilation Claims," United States Congressional Serial Set v2561, 21;

*Lloyd's Register of Shipping*, 99; Chris Maxworthy, personal communication, June 2014; NMDL vessels: *Hope, Olive Branch, Union, Minerva, Lion, Harlequin*; *New York Public Advertiser*, February 2, 1807, 3; *Catalogue of Nantucket Whalers*, 4; *Lloyd's Register of Shipping* (1799), 87; Martin, *Expert Observer*, 289.

39. Stein, ""American Maritime Documents.
40. "Whaling History," NBWM.
41. "NMDL vessel: *Thule*; *Eastern Argus*, June 2, 1845; *Tri-Weekly Argus*, September 3, 1845; *Eastern Argus*, March 17, 1847; Cornell, "Remarks on board Ship Eliza Adams of New Bedford."
42. NMDL vessel: *John Coggeshall*; "Whaling History," NBWM; Martin, *Whalemen and Whaleships*, 56.
43. NMDL vessels: *Amazon, Java II, Chelsea*; "Whaling History," NBWM.
44. NMDL vessels: *Jones, Caledonia, United*; "Whaling History," NBWM.
45. NMDL vessel: *Charles Phelps*; "Whaling History," NBWM; Stearns, *Story of the New England Whalers*, 273–74, 327–29.
46. NMDL vessel: *Pioneer*; "Whaling History," NBWM.
47. NMDL vessel: *Junior*; "Whaling History," NBWM; Stearns, *Story of the New England Whalers*, 341–46.
48. NMDL vessel: *Margaret Scott*; "Whaling History," NBWM; Stearns, *Story of the New England Whalers*, 356.
49. NMDL vessels: *William Hamilton, William Henry*; "Whaling History," NBWM.
50. NMDL vessels: *Addison, Florida II*; Lawrence, *Captain's Best Mate*, 252–53, 254–55, 262–63; Williams, *One Whaling Family*, 206.
51. Pickerill, "American Whalers"; "Whaling History," NBWM; NMDL vessel: *Connecticut*.
52. NMDL vessels: *Atlantic, Java*; "Whaling History," NBWM; Stearns, *Story of the New England Whalers*, 238–82.
53. "Whaling History," NBWM
54. NMDL vessels *Franklin Armadillo, Newark, Golconda, Nye, Martha II*; "Whaling History," NBWM; Burgess, "T.J. Burgess—Form 32"; Megan Pinette, personal communication, March 2019.
55. NMDL vessels: *Ellen A. Swift, John R. Manta, Morning Star*; "Whaling History," NBWM
56. "Benjamin C. Gardner," https://whalinghistory.org; NMDL vessels: *Martha, Nantucket*; Ford, *Pitcairn Island*, 29; Druett, *She Captains*, 165–67.
57. Merill, "Capt. Edward Merrill."
58. "Thomas G. Young," https://whalinghistory.org; "Whaling History," NBWM; NMDL vessels: *Favorite, St. Peter*.

59. "Gersham Leander Cox," https://whalinghistory.org; NMDL Vessels: *Brighton, Magnolia*; "Whaling History," NBWM; Cutter, *Historic Homes and Places*, 1438–39.
60. "James V. Cox," https://whalinghistory.org; NMDL Vessels: *Abigail, Draco.*
61. "George A. Smith," https://whalinghistory.org; NMDL Vessels: *Charles W. Morgan, Triton, Sunbeam.*
62. NMDL vessels: *Hector, Platina, Jireh Perry, Niger*; "Whaling History," NBWM; Martin, *Whalemen and Whaleships*, 56; "Voyages of Amos D. Chase."
63. NMDL vessels: *Contest, Japan*; Owen, Papers of Leander Owen; Everett S. Allen, *Children of the Light*, 214, 230, 239–42, 248–49; Owen, "Archive of Captain Leander Owen."
64. NMDL vessels: *Jireh Perry, Helen Snow, Three Brothers, Coral, Mary and Helen*; "Whaling History," NBWM; Bockstoce, *Whales, Ice & Men*, 168, 207, 210.
65. NMDL vessels: *Belvedere, Daniel Webster, North Star, Rainbow, Thrasher, Susan and Mary*; "Whaling History," NBWM; Owen, "Archive of Captain Leander Owen."; Owen, "Voyages of Owen, Leander C." NMDL: Owen Leander; Bockstoce, *Whales, Ice & Men*, 133, 141, 212, 214.
66. "Christopher Pinkham," https://whalinghistory.org; "Benjamin Worth," https://whalinghistory.org; NMDL vessels *Dolphin, Falmouth, Hector, Falmouth.*
67. "Obed Paddack," https://whalinghistory.org; NMDL vessels *Falkland, Olive Branch, Fortitude*; Headland, *Chronological List*, 88; Memorials of Deceased Friends, 69–72.
68. "Charles F. Coffin," https://whalinghistory.org; NMDL vessel: *Washington.*
69. "Edmund Crowell," https://whalinghistory.org; NMDL vessel: *Clifford Wayne*; Busch, *Whaling Will Never Do for Me*, 60; Wayne and Wady, "Journal of the Clifford Wayne," Tuesday, January 9, 1844 Entry.
70. "James Harvey Shearman, Jr.," https://whalinghistory.org; NMDL vessels: *Young Phenix, Coral, Emigrant.*
71. NMDL vessel: *Lively*; Martin, *Whalemen and Whaleships*, 59; *Eastern Argus*, November 23, 1852.
72. William Baker, *Maritime History of Bath, Maine*, 247; Hallowell and Wilson, "Whaling."
73. "Shoulder Guns."
74. Ibid.
75. Ibid.
76. Barman, "Whatever Happened to the Kanakas?"; Barman, *Maria Mahoi of the Islands.*

## *Chapter 3*

77. Martin, *Whalemen and Whaleships of Maine*, 42; Moment, "Business of Whaling in America," 263; "Whaling Not What It Was," *New York Times*; Headland, *Chronology of Antarctic Exploration*, 169; Glover M. Allen, "Whales and Whaling in New England," 341; Davis, Gallman and Hutchins, "Technology, Productivity, and Profits," 144, 145; NMDL vessel: *Betsey*.
78. NMDL vessels: *Sea Horse, Commerce, Rover*; Davis, Gallman and Hutchins, "Technology, Productivity, and Profits," 149; William Baker, *Maritime History of Bath*, 1080.
79. NMDL vessels: *Charles, Liberty, Mary Ann, Frederick Augustus*.
80. NMDL vessel: *Columbus*; "Whaling History," NBWM.
81. NMDL vessels: *Two Brothers, Martha*; Philbrick, "How Nantucket Came to Be the Whaling Capital."
82. NMDL vessels: *Eliza Barker, Liberty, Robinson Potter, William and Nancy*; *Catalogue of Nantucket Whalers*, 9, 11; William Baker, *Maritime History of Bath*, 1,118; Stearns, *Story of the New England Whalers*, 319–20.
83. NMDL vessels: *General Knox, Nancy, Hero*; Headland, *Chronology of Antarctic Exploration*, 130; Bertrand, *Americans in Antarctica*, 65, 115–16; *Boston Patriot and Daily Mercantile Advertiser*, June 6, 1821.
84. NMDL vessels: *Jason, Mary*; "Whaling History," NBWM; Headland, *Chronology of Antarctic Exploration*, 168, 170; Decker, *Whaling Industry of New London*.
85. NMDL vessels: *George Porter, Harmony*; "Whaling History," NBWM; Headland, *Chronology of Antarctic Exploration*, 135; *Catalogue of Nantucket Whalers*, 12; Starbuck, *History of the American Whale*, 398–99, 452–53.
86. NMDL vessels: *Friendship, Packet*; "Whaling History," NBWM.
87. NMDL vessels: *Bourbon, Liverpool*; "Whaling History," NBWM.
88. NMDL vessels: *Levant, Maine, Sarah Lee*; "Whaling History," NBWM.
89. NMDL vessels: *Mary, Jasper*; "Whaling History," NBWM; Bath Custom Records; Martin, *Whalemen and Whaleships*, 42; William Baker, *Maritime History of Bath*, 290–92, 1093, 1098.
90. NMDL vessels: *General Pike, Laurel, General Gates, Grand Turk*; "Whaling History," NBWM; Bockstoce, *Whales, Ice & Men*, 117; Hughes, "Final Gun"; Baker, *Maritime History of Bath, Maine*, 1,089.
91. NMDL vessels: *Arab, Oxford, Alert, E.R. Sawyer, Lydia*; "Whaling History," NBWM; Headland, *Chronology of Antarctic Exploration*, 151, 156, 187, 190, 192–93; Rogers, "Journal of Erasmus Darwin Rogers," 31–43; Watson,

"Voyage on the Sealer EMELINE," 475–549; Robinson, "Captain J.W. Robinson's Narrative," 301–16; Decker, *Whaling Industry in New London*; Starbuck, *History of American Whale*, 560–61, 576–77; Lipfert, "Maine Whalers"; William Baker, *Maritime History of Bath*, 249, 1,073.

92. NMDL vessels: *Canova, Clarice, Constitution, Franklin, Margaret, Octavia, Iris*; "Whaling History," NBWM.
93. NMDL vessels: *Ann Parry, Gold Hunter, Olive Branch*; "Whaling History," NBWM; Baker, *Maritime History of Bath*, 1,103.
94. NMDL vessel: *Henry Kneeland*; Cloud, *Enoch's Voyage*, 51.
95. Cloud, *Enoch's Voyage*, 161–62.
96. Ibid., 333–34.
97. NMDL vessel: *Henry Kneeland*; "Whaling History," NBWM; Lipfert, "Maine Whalers"; Lawrence, *Captain's Best Mate*, 181, 188.
98. NMDL vessel: *Tamerlane*; "Whaling History," NBWM; Lipfert, "Maine Whalers"; Martin, *Whalemen and Whaleships*, 42; Dow, *Whale Ships and Whaling*, 309; "San Francisco Shanghaiers: 1886–1890"; Lawrence, *Captain's Best Mate*, 41, 158; Hegarty, *Returns of Whaling Vessels*, 38.
99. NMDL vessels: *Adeline, Cassander, Cora*; "Whaling History," NBWM; Lawrence, *Captain's Best Mate*, 135; Stearns, *Story of the New England Whalers*, 363–64.
100. William Baker, *Maritime History of Bath*, 333.
101. NMDL vessel: *Arbella*; William Baker, *Maritime History of Bath*, 231, 333, 340, 1,073.
102. NMDL vessels: *Eagle, McLellan, Pilgrim, Sarah Louisa, Wolga*; "Whaling History," NBWM; William Baker, *Maritime History of Bath*, 386, 1099, 1,119.
103. NMDL vessels: *Amanda, Charles Adams, George Washington, Philetus, Vermont*; "Whaling History," NBWM; Headland, *Chronology of Antarctic Exploration*, 147; A.H. Clark, "Antarctic Fur Seal."
104. NMDL vessel: *Brunette*; Lundeberg, "Samuel Colt's Submarine," 37–38.
105. *National Daily Intelligencer*, April 15, 1844; Lundeberg, "Samuel Colt's Submarine Battery," 44.
106. *National Daily Intelligencer*, April 19, 20, 22, 1844; *Alexandria Gazette and Virginia Advertiser*, April 22, 1844; Peck, *Round Shots to Rockets*, 102–3; Lundeberg, "Samuel Colt's Submarine Battery," ii, 46.
107. NMDL vessels: *Avis, Dwight, Harriet, Brunswick*; Headland, *Chronology List of Antarctic Expeditions*, 172; Parry, "Yankee Whalers in Siberia," 42; Bockstoce, *Whales, Ice & Men*, 103–4; Reynolds and Martin, *Singleness of Purpose*, 176; William Baker, *Maritime History of Bath*, 224, 1,074, 1,090.

108. NMDL vessel: *Coriolanus*; Headland, *Chronology List of Antarctic Expeditions*, 171; Starbuck, *History of the American Whale*, 420–21, 580–81.
109. NMDL vessels: *Emma, Fenelon, Marcia, Majestic*; "Whaling History," NBWM; William Baker, *Maritime History of Bath*, 1,085, 1,098; Lawrence, *Captain's Best Mate*, 90–91, 116, 123.
110. NMDL vessels: *Pleiades, William and Joseph, William Badger, Lucas*; Headland, *Chronology List of Antarctic Expeditions*, 165; Starbuck, *History of the American Whale*, 296–97, 314–15, 354–55, 398–99.
111. NMDL vessels: *Cherokee, Montpelier, Condor, Two Brothers*; Lipfert, "Maine Whalers"; Bockstoce, *Whales, Ice & Men*, 52.

## *Chapter 4*

112. Martin, *Around the World*; Lesley Martin, personal communications, March 9, 2019.
113. NMDL vessel: *Lucy Ann*; Lipfert, "Maine Whalers"; Martin, "Successful Whaling Voyage of the LUCY ANN," 87, 102; William Baker, *Maritime History of Bath, Maine*, 659, 1,097; Martin, *Around the World*; Martin, *Whalemen and Whaleships*, 41–42.
114. NMDL vessel: *Romulus*; Headland, *Chronology List of Antarctic Expeditions*, 178, 187; Starbuck, *History of the American Whale*, 566–67.
115. NMDL vessels: *Emeline, Junius, Jeannette, LeBaron, Mac, Noble, Pantheon, Harvest*; "Whaling History," NBWM; *New York Times*, November 3, 1907; William Baker, *Maritime History of Bath*, 1,093; Headland, *Chronology List of Antarctic Expeditions*, 161; Starbuck, *History of the American Whale*, 342–43, 420–21, 444–45, 568–69; Spence, *Treasures of the Confederate Coast*, 151; Dictionary of American Naval Fighting Ships.
116. NMDL vessels: *Armadillo, Clement, Copia, Lagrange, Malta*; "Whaling History," NBWM.
117. NMDL vessels: *Edward, Geneva, Imogene, Peri, Vermont, Emily Morgan*; "Whaling History," NBWM; Headland, *Chronology List of Antarctic Expeditions*, 170; Dictionary of American Naval Fighting Ships; "Stone Fleet"; Reynolds and Martin, *Singleness of Purpose*, 177–78; Starbuck, *History of the American Whale*, 296–97, 37–73, 440–41; Colby, *For Oil and Buggy Whips*; Lipfert, "Maine Whalers"; *New York Times*, November 7, 1871, 8.
118. William Baker, *Maritime History of Bath*, 258.
119. NMDL vessels: *Roman II, Emma Jane*; "Whaling History," NBWM; Headland, *Chronology List of Antarctic Expeditions*, 153, 156, 199–203; Starbuck,

*History of American Whale*, 550–51, 650–51; William Baker, *Maritime History of Bath*, 258, 1109.

120. NMDL vessel: *Franklin*; "Whaling History," NBWM; Headland, *Chronology List of Antarctic Expeditions*, 188, 200, 203, 206; Starbuck, *History of the American Whale Fishery*, 648–49, 650–51; A.H. Clark, "Antarctic Fur Seal"; Allen, "Fur Seal Hunting," in Jordan *Fur Seals*.

121. NMDL vessel: *Tiger*; Bockstoce, *Whales, Ice & Men*, 181–82.

122. Bockstoce, *Whales, Ice & Men*, 61.

123. NMDL vessel: *Tiger*; Bockstoce, *Whales, Ice & Men*, 94, 130.

124. NMDL vessels: *Bolton, Somerset, Leonidas*; "Whaling History," NBWM; Headland, *Chronology List of Antarctic Expeditions*, 167; Starbuck, *History of the American Whale*, 340–41, 370–71, 416–17, 430–31.

125. NMDL vessels: *Mary and Martha, Venice*; "Whaling History," NBWM.

126. NMDL vessels: *Cossack, North America*; Lipfert, "Maine Whalers"; Bockstoce, *Whales, Ice & Men*, 51; "Whaling History," NBWM.

127. NMDL vessels: *Nahant, Zoroaster, Charleston, Tyleston, Cervantes, White Oak, Stonington*; "Whaling History," NBWM; Headland, *Chronology List of Antarctic Explorations*, 164–65; Decker, *Whaling Industry in New London*; Lipfert, "Maine Whalers"; *Catalogue of Nantucket Whalers*, 42–43, 45; William Baker, *Maritime History of Bath, Maine*, 1,078.

128. NMDL vessels: *Montezuma, General Scott, Laurens, Elizabeth*; Lipfert, "Maine Whalers"; Gray, "Light Airs from the South," 113; Stearns, *Story of the New England Whalers*, 357; Lawrence, *Captain's Best Mate*, 168; "Whaling History," NBWM.

129. NMDL vessels: *Sarah, Millinoket, Tamoree*; "Whaling History," NBWM.

130. NMDL vessels: *Glendower, Globe, Highlander*; Lipfert, "Maine Whalers"; "Whaling History," NBWM.

131. NMDL vessels: *Alice Frazier, Avola, Margaretta, Sea Flower, Warwick*; Lipfert, "Maine Whalers"; "Whaling History," NBWM; Gray, "Light Airs from the South," 113.

132. NMDL vessel: *George Henry*; "Whaling History," NBWM; Hall, *Life with the Esquimaux*, 442.

133. NMDL vessels: *American, Edgar, Helen Augusta, Isabella, J.E. Donnell, Tahmaroo*; Bockstoce, *Whales, Ice & Men*, 117; Spence, *Treasures of the Confederate Coast*, 146; "USS *American*"; "Whaling History," NBWM; Wiliam Baker, *Maritime History of Bath, Maine*, 626, 1,093; "*Tahmaroo* Journal."

134. NMDL vessels: *Germ, Warren, Janet*; "Whaling History," NBWM; Lipfert, "Maine Whalers"; Martin, *Whalemen and Whaleships*, 42; "*Janet* Logbook 1877–1879: Log 843."

135. NMDL vessels: *Afton, Antelope, James Andrews, Osceola III*; "Whaling History," NBWM; Stearns, *Story of the New England Whalers*, 294–95.
136. NMDL vessels: *Henry Trowbridge, Keoka, N.D. Chase, Winthrop, Cleora*; "Whaling History," NBWM.
137. NMDL vessels: *Delaware, Homer, Ionia, Mary Wilder, Oregon*; Lipfert, "Maine Whalers"; "Whaling History," NBWM.
138. NMDL vessels: *Aerial, Amaret, S.H. Waterman, San Francisco*; "Whaling History," NBWM; Lipfert, "Maine Whalers"; Kinney, *Vessels of Way Down East*, 124.
139. NMDL vessels: *Sarah B. Hale, Manuel Ortiz, Walter Irving, Hannah Brewer, Lady Suffolk*; "Whaling History," NBWM; Lipfert, "Maine Whalers"; Kinney, *Vessels of Way Down East*, 124, 128.
140. NMDL vessels: *Civilian, D.M. Hall, Florence, Alfred Gibbs, Helen Snow*; Lipfert, "Maine Whalers"; "Whaling History," NBWM; Baker, *Maritime History of Bath*, 1,088, 1,112.
141. NMDL vessels: *Flying Cloud, Sarah E. Spear, Lively, Hunter*; "Whaling History," NBWM; Lipfert, "Maine Whalers"; Headland, *Chronology List of Antarctic Explorers*, 180; Martin, *Whalemen and Whaleships*, 59; *Eastern Argus*, November 23, 1852; "San Francisco Shanghaiers," Mystic Seaport Museum.
142. Bockstoce, *Whales, Ice & Men*, 65.
143. NMDL vessels: *Robert Morrison, Sea Breeze, Sophia Thornton*; "Whaling History," NBWM.
144. NMDL vessels: *Trinity, Roman, Flying Fish, Lizzie P. Simmons, Golden West*; "Whaling History," NBWM; Headland, *Chronology*, 197–99, 208–9, 213, 215; Decker, *Whaling Industry of New London*; Schultz, *Inventory of the Logbooks*; Boumphrey, *Visit to South Georgia*, 85–92; A.H. Clark, "Antarctic Fur Seal"; Fuller, *Master of Desolation*, 123, 158–59.
145. NMDL vessels: *Pamelia, Rose Pool, Florence, Francis (Frances) Palmer*; "Whaling History," NBWM; "Record of American and Foreign Shipping," Mystic Seaport Museum, 370; "San Francisco Shanghaiers," Mystic Seaport Museum; "Alaska Shipwrecks 1750–1887," Alaska Genealogy Trails; Aldrich, *Eight Months in Arctic*, 8; Bockstoce, "Nineteenth Century Commercial Shipping," 63; Bockstoce, *Whales, Ice & Men*, 193.
146. NMDL vessels: *Wavelet, A. Houghton, Graduate, John Hathaway*; "Whaling History," NBWM; Lipfert, "Maine Whalers"; Kinney, *Vessels of Way Down East*, 134.
147. NMDL vessels: *Kingfisher, Silver Cloud*; "Whaling History"; Lipfert, "Maine Whalers"; Gaines, *Encyclopedia*, 15; Lawrence, *Captain's Best Mate*, 50, 80, 204.

148. NMDL vessels: *Hidalgo, Tempest, Architect, U.D.*; "Whaling History," NBWM; "Record of American and Foreign Shipping," 502; Bockstoce, "Nineteenth Century Commercial Shipping Losses," 67; "San Francisco Shanghaiers," Mystic Seaport Museum; Kinney, *Vessels of Way Down East*, 135.
149. NMDL vessel: *Pilot's Bride*; Headland, *Chronology List of Antarctic Explorers*, 213; Snow, *Main Beam*, 176; Fuller, *Master of Desolation*, xxii, 121, 127, 129; "Report on the Sinking of Schooner Pilot's Bride," Federal Archives and Record Center.
150. Fuller, *Master of Desolation*, 131.
151. NMDL vessel: *Pilot's Bride*; Fuller, *Master of Desolation*, 132, 144–45, 151–52.
152. NMDL vessels: *Pilot's Bride, Trinity*; Fuller, *Master of Desolation*, 123, 158, 164; Headland, *Chronology List of Antarctic Explorers*, 213.
153. Fuller, *Master of Desolation*, 164.
154. Ibid., 165.
155. Ibid., 167.
156. Ibid., 168.
157 NMDL vessels: *Pilot's Bride, Francis Allyn*; Fuller, *Master of Desolation*, 251, 255, 276.
158. NMDL vessel: *John Carver*; "Whaling History," NBWM; Lipfert, "Maine Whalers"; "Record of American and Foreign Shipping," 534; Bockstoce, "Nineteenth Century Commercial Shipping Losses," 65; "Alaska Shipwrecks 1750–1887," Alaska Genealogy Trails; "San Francisco Shanghaiers," Mystic Seaport Museum; Bockstoce, *Whales, Ice & Men*, 150.
159. NMDL vessels: *A.J. Ross, Astoria, D.N. Richards, Eothen, Starlight*; c; Lipfert, "Maine Whalers."
160. NMDL vessel: *Emma F. Herriman*; "San Francisco Shanghaiers," Mysic Seaport Museum; Farr, "Slow Boat to Nowhere," 167; Thompkins, "Black Ahab," 75–84.

## *Chapter 5*

161. Spence, *Treasures of the Confederate Coast*, 142–152, 159–164; Gaines, *Encyclopedia*, 141; "Stone Fleet."
162. Melville, "Stone Fleet."
163. Stearns, *Story of the New England Whalers*, 391–92.
164. NMDL vessels: *American, Leonidas, Noble, Montezuma, Peri, Edward, Marcia, Majestic, Cossack*; Gaines, *Encyclopedia*, 47, 141, 150–52, 197; "Whaling

History," NBWM; *War of the Rebellion* Series 1, Vol. 12: 418, 421, 511; Series 2, Vol. 1: 67, 132, 134, 150, 162; Dictionary of American Naval Fighting Ships; Spence, *Treasures of the Confederate Coast*, 146, 151; Spence, *Shipwrecks of South Carolina and Georgia*, 711; Headland, *Chronology List of Antarctic Expeditions*, 161; Dictionary of American Naval Fighting Ships; Starbuck, *History of the American Whale*, 342–43, 420–21, 444–45, 568–69; "Stone Fleet."

165. NMDL vessel: *Polar Star*; "Whaling History," NBWM; Gaines, *Encyclopedia*, 139; Hearn, *Gray Raiders*, 291.
166. NMDL vessel: *Ebenezer Dodge*; "Whaling History," NBWM; Gaines, *Encyclopedia*, 14; *Official Records of the Union*, 646, 728, 744; Semmes, *Service Afloat*, 278–79.
167. NMDL vessels: *Ocean Rover, Courser, Virginia, Alert, Levi Starbuck, Kate Cory, Lafayette, Nye*; Gaines, *Encyclopedia*, 17–18, 21, 23–24; "Whaling History," NBWM; *Official Records of the Union*, vol. 1, 551, 780, 788–89, 790, 802, vol. 2: 201, 203–4, 685, 740–741; Summersell, *Journal of George Townley Fullam*, 106, 210; Semmes, *Service Afloat*, 431–33, 441, 442, 493, 604, 611–12; Hearn, *Gray Raiders*, 167–68, 169, 194–95; Stearns, *Story of the New England Whalers*, 382.
168. NMDL vessels: *Edward, Pearl, Hector, Edward Carey, Harvest*; "Whaling History," NBWM; Gaines, *Encyclopedia*, 14, 139; Hearn, *Gray Raiders*, 282–83; *Official Records of the Union*, vol. 3, 588, 760, 787, 788–89, 792, 804, 817–19.
169. NMDL vessel: *Abigail*; "Whaling History," NBWM; Gaines, *Encyclopedia*, 20, 139; *Official Records of the Union*, vol. 3, 789, 792, 822–23; Hearn, *Gray Raiders*, 284–85.
170. NMDL vessel: *Sophia Thornton*; Lipfert, "Maine Whalers"; Headland, *Chronology List of Antarctic Explorers*, 177; Gaines, *Encyclopedia*, 21; *Official Records of the Union*, vol. 3, 790, 792, 826–27.
171. Quoted in Bockstoce, *Whales, Ice & Men*, 115.
172. NMDL vessels: *Euphrates, Jireh Swift, Milo*; "Whaling History," NBWM; Gaines, *Encyclopedia*, 20; *Official Records of the Union*, vol. 3: 790, 792, 825, 826; Hearn, *Gray Raider*, 287; Stearns, *The Story of the New England Whalers*, 385.
173. NMDL vessels: *Susan and Abigail, William Thompson, General Williams*; "Whaling History," NBWM; Gaines, *Encyclopedia*, 20, 21; *Official Records of the Union*, vol. 3, 790, 792, 825, 827.
174. NMDL vessels: *Nimrod, Ocean Rover, Catharine, Gypsy, William C. Nye, General Pike*; "Whaling History," NBWM; Gaines, *Encyclopedia*, 20; *Official Records*, vol. 3, 790, 791–92, 828–29; Hearn, *Gray Raiders*, 288, 289–90;

Headland, *Chronology List of Antarctic Explorers*, 172; Parry, "Yankee Whalers in Siberia," 42; Bockstoce, *Whales, Ice & Men*, 103–4, 117.

175. NMDL vessels: *Congress, Favorite, Hillman, Martha, Nassau*; "Whaling History," NBWM.

176. NMDL vessel: *Waverly*; "Whaling History," NBWM; Gaines, *Encyclopedia*, 20–21; Hearn, *Gray Raiders*, 290–91; *Official Records of the Union*, vol. 3, 791–92, 828–29; Reynolds and Martin, *Singleness of Purpose*, 176.

177. NMDL vessels *James Maury, Nile*; "Whaling History," NBWM.

## Chapter 6

178. NMDL vessels: *William Grozier, J. Taylor, Charles Thompson, Benjamin T. Crocker*; Lipfert, "Maine Whalers."

179. NMDL vessels: *Thomas Hunt, John Atwood, Willie Irving, Sullivan, Augustine Kobbe.*

180. NMDL vessels: *John H. Perry, Delia Hodgkins, Mary E. Higgins*; Lipfert, "Maine Whalers"; William Baker, *Maritime History of Bath*, 514, 1,091, 1,106.

181. "Overview of American Whaling," NBWM.

182. "Whale Fishery in 1870," *New York Times.*

183. NMDL vessels: *Florence, Minerva*; "American Lloyd's Register," Mystic Seaport Museum, 64; Bockstoce, "Nineteenth Century Commercial Shipping," 62; *New York Times*, November 6, 1871.

184. Walkerman, "Captain's Nightmare," 1.

185. *Friend*, November 1, 1871, reprinted in Mrantz, *Whaling Days in Old Hawaii*, 36.

186. Ibid.

187. NMDL vessel: *Emily Morgan*; Lipfert, "Maine Whalers"; *New York Times*, November 7, 1871; Everett S. Allen, *Children of the Light*, 213; Bockstoce, *Whales, Ice & Men*, 151; "Whaling History," NBWM; Stearns, *Story of the New England Whalers*, 410–14.

188. Everett S. Allen, *Children of the Light*, 220.

189. Ibid., 230.

190. "We Left Not a Moment Too Soon," *Honolulu Friend*; Walkerman, "Captain's Nightmare," 34.

191. Quoted in Bockstoce, *Whales, Ice & Men*, 156.

192. Walkerman, "Captain's Nightmare," 39, 93; "Log of Bark EMILY MORGAN," quoted in Everett S. Allen, *Children of the Light*, 247.

193. NMDL vessels: *Europa, Contest*; "Whaling History," NBWM; Walkerman, "Captain's Nightmare," 42, 70; Everett S. Allen, *Children of the Light*, 236, 237, 252; Bockstoce, *Whales, Ice & Men*, 157–58, 163, 207.
194. Walkerman, "Captain's Nightmare," 1; Mrantz, *Whaling Days in Old Hawaii*, 37; Mishkar, "Arctic Whaling Ships Discovered," 1; Everett S. Allen, *Children of the Light*, 263; Bockstoce, *Whales, Ice & Men*, 161–62.
195. NMDL vessels: *Florence, Helen Snow*; Lipfert, "Maine Whalers"; *New York Times*, October 13, 1872; Bockstoce, *Whales, Ice & Men*, 168, 170.
196. *New York Times*, November 29, 1871; April 19, 1880.
197. Ibid., January 3, 1873; November 14, 1871; July 19, 1891; "Whaling Disaster of 1871"; A.H. Clark, "Scientists Find Wreckage; Stewardson, "Whaling Shipwreck Site"; Mishkar, "Arctic Whaling Ships Discovered," 2.
198. NMDL vessels: *Florence, Three Brothers*; *New York Times*, October 13, 1874; Mrantz, *Whaling Days in Old Hawaii*, 37; "Twelve Ships Lost at Sea," *New York Times*; "Abandoned Whalers," *New York Times*.
199. "Twelve Ships Lost at Sea," *New York Times*.
200. *New York Times*, October 23, 1876; *New York Times*, "Effects of the Arctic Disaster."
201. Martin, "Bath-Built Steam Whalers," 46.
202. *Daily Times*, June 24, 1874; William Baker, *Maritime History of Bath*, 603.
203. McCauley, "Vessel *Fresno* Burns"; Singer, "Whaling Ship"; William Baker, *Maritime History of Bath*, 603, 1,087.
204. NMDL vessels: *Leonora, Adelia Chase*; "Whaling History," NBWM; Headland, *Chronology List of Antarctic Explorers*, 212; Sherman, *Whaling Logbooks and Journals*; Balch, "Antarctica Addenda," 81–88; William Baker, *Maritime History of Bath*, 1,079.
205. NMDL vessel: *John Winthrop*; "Whaling History," NBWM; "San Francisco Shanghaiers," Mystic Seaport Museum; Bockstoce and Batchelder, "Chronological List," 85, 86; Bockstoce, *Whales, Ice & Men*, 280–82, 338; William Baker, *Maritime History of Bath*, 1,094.
206. NMDL vessel: *Sarah W. Hunt*; "Whaling History," NBWM; William Baker, *Maritime History of Bath*, 1,092.
207. NMDL vessels: *James Allen, James Allen II, Josephine*; "Whaling History," NBWM; Lipfert, "Maine Whalers"; "San Francisco Shanghaiers," Mystic Seaport Museum; Martin, *Whalemen and Whaleships*, 43; Hegarty, *Returns of Whaling Vessels*, 17, 23, 31, 39, 40; William Baker, *Maritime History of Bath*, 1,072, 1,094.

208. NMDL vessel: *Lucretia*; Lipfert, "Maine Whalers"; "San Francisco Shanghaiers," Mystic Seaport Museum; Bockstoce and Botkin, "Harvest of Pacific Walruses," 185; Bockstoce, *Whales, Ice & Men*, 212, 257, 309; William Baker, *Maritime History of Bath*, 1,097.

209. NMDL vessels: *Mabel, George and Susan, Attleboro, Cameo*; "Whaling History," NBWM; Baker, *Maritime History of Bath*, 663, 1,074, 1,077, 1,097; Lipfert, "Maine Whalers"; Bockstoce, *Whales, Ice & Men*, 300–1.

210. NMDL vessel: *Mary and Helen*; "Whaling History," NBWM; Lipfert, "Maine Whalers"; Lundberg, "Thar She Blows!"; Stewardson, "Loss of the Navarch"; *New York Times*, October 12, 1880; Dorsett, "Maine Whalers," 39; Martin, *Whalemen and Whaleships*, 46–47; Martin, "Bath-Built Steam Whalers," 46; William Baker, *Maritime History of Bath*, 533–34, 663, 1,098; Stearns, *Story of the New England Whalers*, 239–40.

211. NMDL vessel: *Wanderer*; "Whaling History," NBWM.

212. NMDL vessel: *Belvedere*; "Whaling History," NBWM; Bockstoce and Batchelder, "Chronological List," 86, 87, 89; "San Francisco Shanghaiers," Mystic Seaport Museum; Baker, *Maritime History of Bath*, 1,075; Lipfert, "Maine Whalers"; Harrison, *Nome and Seward Peninsula*, 17–18.

213. NMDL vessels: *Belvedere, Jesse H. Freeman, Orca*; Bockstoce and Batchelder, "Chronological List," 86, 87, 89; Lundberg, "Dr. Samuel J. Call"; *New York Times*, September 7, 1898; Bockstoce, *Whales, Ice & Men*, 296, 338.

214. "Whaling History," NBWM; NMDL vessels: *Mary and Helen II, Navarch, Young Phoenix, Montezuma*; Lipfert, "Maine Whalers"; "San Francisco Shanghaiers," Mystic Seaport Museum; Bockstoce and Batchelder, "Chronological List," 85, 86, 87, 88, 89; Arthur James Allen, *Whaler and Trader*, 29; Stone, "Whalers and Missionaries," 104, 114, 115; Bockstoce, *Whales, Ice & Men*, 101, 282, 286, 306, 327, 329.

215. NMDL vessel: *Beluga*; Lipfert, "Maine Whalers"; Bockstoce and Batchelder, "Chronological List," 85, 86, 87, 88, 89; Bryant, "Drift Casks in the Arctic Ocean," 158; *New York Times*, November 3, 1907; Bockstoce, *Whales, Ice & Men*, 101, 282, 286, 306, 327, 329.

216. NMDL vessel: *North Star*; Lipfert, "Maine Whalers"; "Alaska Shipwrecks 1750–1887," Alaska Genealogy Trails; "Record of American and Foreign Shipping," 763; Bockstoce, "Nineteenth Century Commercial Shipping," 64; Bockstoce, *Whales, Ice & Men*, 212, 282; William Baker, *Maritime History of Bath*, 1,103; Martin, "Bath-Built Steam Whalers," 78.

217. NMDL vessels: *Thrasher, Navarch*; Lipfert, "Maine Whalers"; "San Francisco Shanghaiers," Mystic Seaport Muesum; Bockstoce and Batchelder, "Chronological List," 85, 90; Bryant, "Drift Casks in the

Arctic Ocean," 157; "Ill-Fated Navarch," *New York Times*; *New York Times*, November 3, 1907; Bockstoce, *Whales, Ice & Men*, 257; Martin, "Bath-Built Steam Whalers," 47; William Baker, *Maritime History of Bath*, 1,115.

218. NMDL vessel: *Jesse H. Freeman*; "San Francisco Shanghaiers," Mystic Seaport Museum; Bockstoce, "Nineteenth Century Commercial Shipping," 67; Bockstoce and Batchelder, "Chronological List," 85, 86, 87; Stone, "Whalers and Missionaries at Herschel Island," 114; Lundberg, "Dr. Samuel J. Call," 5; Bockstoce, *Whales, Ice & Men*, 257, 292, 296; William Baker, *Maritime History of Bath*, 621, 1,087.

219. Bockstoce, *Whales, Ice & Men*, 298.

220. NMDL vessel: *Bertha D. Nickerson*; "Whaling History," NBWM.

221. NMDL vessels: *Morning Star, Herman*; "Whaling History"; *New York Times*, November 3, 1907; Martin, *Whalemen and Whaleships*, 51–52; William Baker, *Maritime History of Bath*, 1,101; "San Francisco Shanghaiers," Mystic Seaport Museum.

222. NMDL vessels: *William Baylies, Bowhead*; Lipfert, "Maine Whalers"; "San Francisco Shanghaiers," Mystic Seaport Museum; Bockstoce and Batchelder, "Chronological List," 86; "News of Ice-Bound Whalers," *New York Times*; *New York Times*, November 3, 1907; Martin, *Whalemen and Whaleships*, 49; Martin, "Bath-Built Steam Whalers," 83, 89; William Baker, *Maritime History of Bath*, 1,075.

223. NMDL vessels: *Jane Gray, Eliza, Bounding Billow, Andrew Hicks*; "San Francisco Shanghaiers," Mystic Seaport Museum; Bockstoce, *Whales, Ice & Men*, 306; William Baker, *Maritime History of Bath*, 1,089.

224. NMDL vessel: *William Lewis*; "Whaling History," NBWM; Lipfert, "Maine Whalers"; "Record of American and Foreign Shipping," 953; Bockstoce, "Nineteenth Century Commercial Shipping," 66; "San Francisco Shanghaiers," Mystic Seaport Museum; Martin, *Whalemen and Whaleships*, 49; William Baker, *Maritime History of Bath*, 627, 1,096.

225. NMDL vessels: *Navarch, Karluk*; Stewardson, "Loss of the Navarch"; Bockstoce, *Whales, Ice & Men*, 290; Brower, *Fifty Years Below Zero*, 183–84.

226. Stewardson, "Loss of the Navarch."

227. Brower, *Fifty Years Below Zero*, 185; Stewardson, "Loss of the Navarch."

228. Brower, *Fifty Years Below Zero*, 186; and Stewardson, "Loss of the Navarch."

229. Brower, *Fifty Years Below Zero*, 187, 190; Stewardson, "Loss of the Navarch"; Bockstoce, *Whales, Ice & Men*, 291–92.

230. Brower, *Fifty Years Below Zero*, 193.

231. NMDL vessels: *Navarch, Thrasher*; "Ill-Fated Navarch," *New York Times*; "Disaster, Desertion and Death," *San Francisco Call*; Bockstoce, *Whales, Ice & Men*, 292; Brower, *Fifty Years Below Zero*, 197–98, 199; "Escape From an Ice Pack," *New York Times*; "Terrible Disaster to a Whaling Vessel," *Bay of Plenty Times*; "News of Ice-Bound Whalers," *New York Times*; Stewardson, "Loss of the Navarch."

232. NMDL vessels: *Navarch, Newport, Fearless*; "Guttner et al.," 618, 619–20, 623; Bockstoce, *Whales, Ice & Men*, 319.

233. NMDL vessel: *Navarch*; "Whaling History," NBWM; Lipfert, "Maine Whalers"; Bockstoce, "Nineteenth Century Commercial Shipping," 67; Bockstoce and Batchelder, "Chronological List," 85, 86; Stewardson, "Loss of the Navarch"; William Baker, *Maritime History of Bath*, 627, 1,102; Lundberg, "Dr. Samuel J. Call"; Boyd, "Jarvis and the Alaskan Reindeer," 76; "Modern Scrimshaw By Artist Salm Rashidi."

234. NMDL vessels: *Valkyria, Viola*; Lipfert, "Maine Whalers."

235. Glover M. Allen, "Whales and Whaling in New England," 341.

## *Chapter 7*

236. Kushner, "Hellships," 81, 82–83, 90; Bockstoce and Botkin, "Harvest of Pacific Walruses," 183, 184; Bockstoce, *Whales, Ice & Men*, 129–30.

237. Moment, "Business of Whaling," 267, 281; Glover M. Allen, "Whales and Whaling in New England," 341; Farr, "Slow Boat to Nowhere," 166; *New York Times*, March 3, 1878; "Last of the New Bedford Whalers," *New York Times*; *New York Times*, August 5, 1879.

238. *New York Times*, November 25, 1883; "Whaling Not What It Was," *New York Times*; Mrantz, *Whaling Days*, 35; Walkerman, "Captain's Nightmare," 4–5, 8, 12; "In Search of the Lost Whaling Fleets"; Boyd, "Jarvis and the Alaskan Reindeer," 74–82; Lundberg, "Thar She Blows!"; Bertholf, "Rescue of the Whalers"; "Incredible Alaska Overland Rescue"; Bockstoce and Batchelder, "Chronological List," 81; Stewardson, "Arctic Whalemen Pushed Limits"; "Last of the New Bedford Whalers," *New York Times*; "Whaling by Airplanes," *New York Times*.

239. Lagerbom, "Fate of *Louise*," 301–11.

240. Snow, *Main Beam*, 448; "Paloverde"; "Snow Shipyards, Rockland ME"; Priolo, "ATA-215"; "ATA-215" Dictionary of American Naval Fighting Ships; Ronne and Liss, *Ronne Expedition to Antarctica*, 20; Edith Ronne, *Transcript of Oral Interview—July 20, 2000*, 10; Darlington, *My Antarctic*

*Honeymoon*, 70–71; Robert H.T. Dodson, personal communication, December 10, 2010.

241. Dodson, Transcript of Oral Interview, 16.
242. Finn Ronne, *Antarctic Conquest*, 50–51, 274–75; Darlington, *My Antarctic Honeymoon*, 16, 67, 124–25; Edith Ronne, *Antarctica's First Lady*, 41; "Robert H.T. Dodson, Oral Interview," 18, 31.
243. "Robert H.T. Dodson, Oral Interview," 62.
244. Morton, "New Chore"; Snow, *Main Beam*, 447–48, 498; "Noble Fleet of Sealers."
245. Morton, "New Chore"; Fitzgerald, "Operation Look See"; Snow, *Main Beam*, 447–48, 498; Friends of Hydrography, "Peter Dodge—Biographical Notes"; Friends of Hydrography, "Brown, Earl"; Friends of Hydrography, "Melanson, Russell Cameron."
246. Friends of Hydrography, "Yeaton, George M."; "Sealing Ship Sinks," *Times*; Snow, *Main Beam*, 412; Robert H.T. Dodson, personal communication.
247. "Whaling History," NBWM.
248. Stearns, *Story of the New England Whalers*, 417–18.

# BIBLIOGRAPHY

"Alaska Shipwrecks 1750–1887." Alaska Genealogy Trails. http://genealogytrails.com/alaska/shipwreck_list.htm.

Aldrich, Herbert L. *Eight Months in Arctic and Siberia with the Arctic Whalemen.* New Bedford, MA: New Bedford Whaling Museum, 1937.

Allen, Arthur James. *A Whaler and Trader in the Arctic.* Anchorage: Alaska Northwest Publishing Co., 1978.

Allen, Everett S. *Children of the Light.* New York: Little, Brown and Co., 1973.

Allen, Everett S. *Children of the Light: The Rise and Fall of New Bedford Whaling and the Death of the Arctic Fleet.* Orleans, MA: Parnassus Imprints, 1973.

Allen, Glover M. "The Whalebone Whales of New England." *Memoirs of the Boston Society of Natural History* 8, no. 2. Monograph. Boston: BSNH, 1916.

———. "Whales and Whaling in New England." *Scientific Monthly* 27, no. 4 (October 1928): 340–43.

Allen, J.A. "Fur Seal Hunting in the Southern Hemisphere." In *The Fur Seals and the Fur Seal Islands of the North Pacific Ocean*, edited by J.S. Jordan. Washington, D.C.: GPO, 1899.

"American Lloyd's Register of American and Foreign Shipping." Mystic Seaport Museum. https://research.mysticseaport.org/item/l0237571883.

"American Offshore Whaling Voyages." National Maritime Digital Library. http://nmdl.org.

"ATA-215." Dictionary of American Naval Fighting Ships. https://www.history.navy.mil/research/histories/ship-histories/danfs.html.

Baker, Sylvanus. Whaling History. https://whalinghistory.org/?s=AE077181023.

Baker, William A. *A Maritime History of Bath, Maine, and the Kennebec River Region*. Bath, ME: Marine Research Society of Bath, 1973.

Balch, E.S. "Antarctica Addenda." *Journal of the Franklin Institute* 157 no. 2 (1904): 81–88.

Barman, Jean. *Maria Mahoi of the Islands*. Vancouver, BC: New Star Books, 2004.

———. "Whatever Happened to the Kanakas? They're Alive and Well in British Columbia." *Beaver* 77 no. 6 (December 1997/January 1988): 12–19.

Bath Custom Records from Robert B. Applebee Collection. Maine Maritime Museum.

*Bay of Plenty Times*. "Terrible Disaster to a Whaling Vessel in the Arctic Ocean." October 20, 1897.

Bertholf, Lt. Ellsworth P. "The Rescue of the Whalers: A Sled Journey of 1,600 Miles in the Arctic Regions." *Harper's New Monthly Magazine*, June 1899.

Bertrand, Kenneth J. *Americans in Antarctica 1775–1948*. New York: American Geographical Society, 1971.

Biscoe, Mark Wyman. *No Pluckier Set of Men Anywhere: The Story of Ships and Men in Damariscotta and Newcastle, Maine*. Newcastle, ME: Lincoln County Publishing Co., 2001.

"Blubber Knife." Maine Memory Network. https://www.mainememory.net/artifact/11488.

Bockstoce, John R. "The Consumption of Caribou by Whalemen at Herschel Island, Yukon Territory, 1890 to 1908." *Arctic and Alpine Research* 12, no. 3 (1980): 381–84.

———. "Nineteenth Century Commercial Shipping Losses in the Northern Bering Sea, Chukchi Sea, and Beaufort Sea." *Northern Mariner* 16, no. 2 (April 2006): 53–68.

———. *Whales, Ice & Men: The History of Whaling in the Western Arctic*. Seattle: University of Washington Press, 1995.

Bockstoce, John R., and Charles F. Batchelder. "A Chronological List of Commercial Wintering Voyages to the Bering Strait Region and Western Arctic of North America: 1850–1910." *American Neptune* 38, no. 2 (1978): 81–91.

Bockstoce, John R., and Daniel B. Botkin. "The Harvest of Pacific Walruses by the Pelagic Whaling Industry, 1848 to 1914." *Arctic and Alpine Research* 14, no. 3 (1982): 183–88.

Boumphrey, R.S., trans. *A Visit to South Georgia by H.W. Klutschak, 1877. British Antarctic Survey Bulletin* #12. London: British Antarctic Society, 1967.

Boyd, William L. "Jarvis and the Alaskan Reindeer Caper." *Arctic* 25, no. 2 (June 1972): 74–82.

Brower, Charles D. *Fifty Years Below Zero: A Lifetime of Adventure in the Far North.* New York: Dodd, Mead & Co., 1942.

Bryant, Henry G. "Drift Casks in the Arctic Ocean." *Proceedings of the American Philosophical Society* 41, no. 169 (April 1902): 154–61.

Bunting, W.H. Bunting. *Live Yankees: The Sewalls and Their Ships*. Bath: Maine Maritime Museum, 2009.

Burgess, Thomas Jefferson. "T.J. Burgess—Form 32—Certificate of Exemption for a Drafted Person on Account of Disability, August 14, 1863." Belfast Historical Society and Museum.

Busch, Briton Cooper. *Whaling Will Never Do for Me: The American Whaleman in the Nineteenth Century*. Lexington: University Press of Kentucky, 2015.

"Case Containing Whale Tooth." Andrew Carnegie Archives. www.accesspadr.org.

*Catalogue of Nantucket Whalers and Their Voyages from 1815 to 1870*. Nantucket, MA: Hussey & Robinson, 1876.

Chase, Amos D. "Voyages of Amos D. Chase." National Maritime Digital Library. http://nmdl.org.

Chase, Fannie. *Wiscasset in Pownalborough: A History of the Shire Town and the Salient Historical Features of the Territory Between the Sheepscot and Kennebec Rivers.* Wiscasset, ME: Anthoensen Press, 1967.

Clark, A.H. "The Antarctic Fur Seal and Sea Elephant Industries." In *The Fisheries and Fishing Industries of the United States*, edited by G.B. Goode, 400–67. N.p., 1887.

Clark, Maureen. "Scientists Find Wreckage of Whaling Ships Trapped When Chukchi Sea off Alaska Froze." *Seattle Times*. October 11, 1998.

Clifford Wayne, Wady, John. "Journal of the Clifford Wayne (Ship) out of Fairhaven, MA, Mastered by Edmund Crowell and John H. Wady and Kept by John H. Wady, on Whaling Voyages between 1841 and 1851." https://archive.org/details/journalofcliffor00clif/page/154.

Cloud, Enoch Carter. *Enoch's Voyage: Life on a Whale Ship 1851–1854.* Wakefield, RI: Moyer Bell, 1994.

Colby, B.L. *For Oil and Buggy Whips*. Mystic, CT: Mystic Seaport Museum, 1990.

Cornell, Weston M. "Remarks on Board Ship ELIZA ADAMS of New Bedford." Journal kept aboard Bark RHINE of New Bedford, 1846–1848. No. 394, Kendall Whaling Museum.

Cutler, Carl C. *Queens of the Western Ocean: The Story of America's Mail and Passenger Sailing Lines*. Annapolis, MD: USNI Press, 1961.

Cutter, William Richard. *Historic Homes and Places and Genealogical and Personal Memoirs Relating to the Families of Middlesex County, Massachusetts*. Vol. 4. New York: Lewis Historical Publishing Co., 1908.

Darlington, Jennie. *My Antarctic Honeymoon*. New York: Doubleday & Co., 1956.

Davis, L.E., R.E. Gallman and T.D. Hutchins. "Technology, Productivity, and Profits: British-American Whaling Competition in the North Atlantic, 1816–1842." *Oxford Economic Papers* 39, no. 4 (December 1987): 738–59.

Day, Jane. "Gil Whitman's Anvil Art." *Down East*, April 1972.

Decker, R.O. *Whaling Industry of New London.* York, PA: G. Schumway, 1986.

Densmore, David C. *The Halo: An Autobiography*. Boston: Voice of Angels Publishing House, 1876.

Dictionary of American Naval Fighting Ships. http://www.history.navy.mil/danfs/c14/cossack.htm.

Dodson, Robert H.T. Transcript of Oral Interview—July 24, 2000. Polar Archival Program. Byrd Polar Research Institute. Ohio State University.

Dorsett, E. Lee. "The Maine Whalers." *Down East*, April 1956.

Dow, George Francis. *Whale Ships and Whaling During Three Centuries, with an Account of the Whale Fishery in Colonial New England*. Salem, MA: Marine Research Society, 1925.

Druett, Joan. *She Captains: Heroines and Hellions of the Sea*. New York: Simon & Schuster, 2001.

Farr, James. "A Slow Boat to Nowhere: The Multi-Racial Crews of the American Whaling Industry." *Journal of Negro History* 68, no. 2 (Spring 1983): 159–70.

Field, Rachel. *God's Pocket: The Story of Captain Samuel Hadlock, Junior of Cranberry Isles, Maine*. New York: Macmillan Co., 1934.

Fitzgerald, W.F. "Operation Look See, Greenland, Fall 1957." *Sikorsky News*. September 1958.

Ford, Herbert. *Pitcairn Island as a Port of Call: A Record, 1790–2010.* Jefferson, NC: MacFarland & Co., 2012.

"French Spoilation Claims." United States Congressional Serial Set v2561. 1889.

Friends of Hydrography. "Bolton, Mike." http://fohcan.org/people/b.html.

———. "Brown, Earl." http://fohcan.org/people/b.html.

———. "Peter Dodge—Biographical Notes." http://fohcan.org.

———. "Yeaton, George M." http://fohcan.org.

Fuller, Joseph J. *Master of Desolation: The Reminiscences of Capt. Joseph J. Fuller.* Edited by Briton Cooper Busch. Mystic, CT: Mystic Seaport Museum Inc., 1980.

Gaines, W. Craig. *Encyclopedia of Civil War Shipwrecks*. Baton Rouge: Louisiana State University Press, 2008.

Gifford, Phylander. *Remarks on Board Ship ADDISON*. December 30, 1839. No. 5, Kendall Whaling Museum, Sharon, MA.

Goode, George Brown. *The Fisheries and Fishery Industries of the United States*. Quoted in William N. Peterson. "'Bony-Fish': The Menhaden Fishery at Mystic, Connecticut." *Log of Mystic Seaport* 33 (1981): 26.

Gray, Alastair C. "'Light Airs from the South': Whalers' Logs in Pacific History." *Journal of Pacific History* 35, no. 1 (June 2000): 109–13.

"Guttner et al. vs Pacific Steam Whaling Co." *Federal Reporter* 96 (September–November 1899): 617–23.

"Hairpin, circa 1840." Maine Memory Network. https://www.mainememory.net/artifact/48948.

Hall, Charles Francis. *Life with the Esquimaux: A Narrative of Arctic Experience in Search of Survivors of Sir John Franklin's Expedition.* Rutland, VT: Charles E. Tuttle & Co., 1970.

Hallowell, Samantha, and Annie Wilson. "Historic Hallowell: Whaling." Maine Memory Network. http://historichallowell.mainememory.net/page/2476/display.html.

Harrison, E.S. *Nome and Seward Peninsula: History, Description, Biographies and Stories*. Seattle, WA: E.S. Harrison, 1905.

Hawes, Charles Boardman. *Whaling*. Garden City, NY: Doubleday, Page & Co., 1924.

Headland, Robert K. *Chronological List of Antarctic Expeditions and Related Historical Events*. Cambridge: Cambridge University Press, 1989.

———. *A Chronology of Antarctic Exploration: A Synopsis of Events and Activities from the Earliest Times Until the International Polar Years, 2007–09*. London: Quaritch, 2009.

Hearn, Chester G. *Gray Raiders of the Sea: How Eight Confederate Warships Destroyed the Union's High Seas Commerce.* Baton Rouge: Louisiana State University Press, 1996.

Hegarty, Reginald B., ed. *Returns of Whaling Vessels Sailing from American Ports: A Continuation of Alexander Starbuck's "History of the American Whale Fishery" 1876–1928*. New Bedford, MA: Old Dartmouth Historical Society and Whaling Museum, 1959.

"History of Captain John O. Morse and His House." https://captainmorsehouse.com/about/history.

*Honolulu Friend*. "We Left Not a Moment Too Soon." November 1, 1871.

Hughes, Dwight. "The Final Gun." Emerging Civil War.August 6, 2015. https://emergingcivilwar.com/2015/08/06/the-final-gun/#_ftn1.

Humiston, Fred. *Blue Water Men—And Women*. Portland, ME: Guy Gannett Publishing, 1965.

"The Incredible Alaska Overland Rescue." Naval Historical Center. www.ibiblio.org/hyerwar/NHC/Alaska-overland-rescue.htm.

"In Search of the Lost Whaling Fleets." National Marine Sanctuaries. http://sanctuaries.noaa.gov/whalingfleet/fatal_blow.html.

"*Janet* Logbook 1877–1879: Log 843." Mystic Seaport. http://library.mysticseaport.org/initiative/PageImage.cfm?BibID=36043.

*A Journal of a Voige to the Pacific Ocean Ship C. MITCHELL*. April 2, 1836. No. 50, Kendall Whaling Museum, Sharon, MA.

Kinney, Joyce E. *The Vessels of Way Down East*. Bangor, ME: Joyce E. Kinney, 1989.

Kushner, Howard I. "'Hellships': Yankee Whaling Along the Coasts of Russian-America, 1835–1852." *New England Quarterly* 45, no. 1 (March 1972): 81–95.

Lagerbom, Charles H. "The Fate of Louise: A Maine-Built 'Down-Easter' at Grytviken Harbor, South Georgia Island." *Mariner's Mirror* 98, no. 3 (August 2012): 301–11.

Laing, Alexander. *Seafaring America*. New York: McGraw-Hill, 1974.

Larkin, Alice True. "Luther Maddocks' Fabulous Whale." *Down East*, November 1977.

Lawrence, Mary Chipman. *The Captain's Best Mate: The Journal of Mary Chipman Lawrence on the Whaler ADDISON, 1856–1860*. Edited by Stanton Garner. Hanover, NH: University Press of New England, 1966.

Leland, Charles. *Algonquin Legends of New England or Myths and Folklore of the Micmac, Passamaquoddy and Penobscot Tribes*. Boston: Houghton, Mifflin & Co., 1884.

Lipfert, Nathan. "Maine Whalers." Maine Maritime Museum, Bath, ME.

"List of Vessels." *Hispanic American Historical Review* 26, no. 4 (November 1946): 602–3.

*Lloyd's Register of Shipping*. London: Lloyds, 1799–1802.

Lundberg, Murray. "Dr. Samuel J. Call in the Arctic Seas." *ExploreNorth* (blog). http://explorenorth.com/library/yafeatures/bl-DrCall.htm.

———. "Thar She Blows! Whaling in Alaska and the Yukon." *ExploreNorth* (blog). http://explorenorth.com/library/yafeatures/blwhaling.htm.

Lundeberg, Philip K. "Samuel Colt's Submarine Battery: The Secret and the

Enigma." *Smithsonian Studies in History and Technology* no. 29 (1974): 1–90.

Maddocks, Luther. *Looking Backward: Memories from the Life of Luther Maddocks.* N.p., 1926.

Martin, John. *Expert Observer, 1866.* Portland, ME: Maine Historical Society, 1866.

Martin, John F. *Around the World in Search of Whales: A Journal of the Lucy Ann Voyage, 1841–1844.* Edited by Kenneth R. Martin. New Bedford, MA: New Bedford Whaling Museum, 2016.

——. "The Bath-Built Steam Whalers." *Down East,* January 1973.

———. "Spirit Voices and a Bath Whaler." *Down East*, November 1973.

———. "The Successful Whaling Voyage of the LUCY ANN of Wilmington, 1837–1839" *Delaware History* 15, no. 2 (October 1972): 152–70.

———. "Voyages of Maine Whaleships." *Down* East, May 1972.

———. *Whalemen and Whaleships of Maine: A Publication of the Marine Research Society of Bath, Maine.* Brunswick, ME: Harpswell Press, 1975.

McCauley, Matt. "Vessel *Fresno* burns in Meydenbauer Bay on April 4, 1923." Free Encyclopedia of Washington State History. http://www.historylink.org/File/4256.

Melville, Herman. "The Stone Fleet." http://www.onlineliterature.com/melville/882.

Memorials of Deceased Friends: Of New England Yearly Meeting, 1841.

Merill, Edward. "Capt. Edward Merrill: Ship Captain, Inventor, Developer & Artist." https://www.merrillswaterfront.com/captain-merrill.

Mishkar, Larry. "Arctic Whaling Ships Discovered." *Field Notes.* July 31, 2008.

"Modern Scrimshaw by Artist Salm Rashidi (1990s) Depicting the Whaling Bark Navarch 1892–7." http://www.icollector.com/Modern-scrimshaw-by-artist-Salman-Rashidi-1990s-depicting-the-whaling-bark-Navarch-1892-7_i8480382

Moment, David. "The Business of Whaling in America in the 1850s." *Business History Review* 31, no. 3 (Autumn 1957): 261–91.

Morton, Jackson L., "A New Chore for the Arctic Sealer." *Ships and the Sea* (Spring 1958): 39–40.

Mrantz, Maxine. *Whaling Days in Old Hawaii.* Honolulu, HI: Aloha Graphics and Sales, 1976.

"National Archives Project." In *Ship Registers of New Bedford, Massachusetts.* Boston: National Archives Project, 1940.

Neil, George F. *Journal Kept Aboard the Ship ELBE.* Barbara E. Johnson Whaling Collection, Princetown, NJ.

*New York Times.* "The Abandoned Whalers." November 20, 1876.

———. "Effects of the Arctic Disaster." October 25, 1876.
———. "Escape from an Ice Pack." September 15, 1897.
———. "The Ill-Fated Navarch." October 10, 1897.
———. "Last of the New Bedford Whalers: The Wreck of the Wanderer and the Vanished Great Days of Seamanship Celebrated by Herman Melville." September 14, 1924.
———. "News of Ice-Bound Whalers." December 2, 1897.
———. "Twelve Ships Lost at Sea." October 22, 1876.
———. "The Whale Fishery in 1870." February 6, 1871.
———. "Whaling by Airplane Shocks Old Timers." May 4, 1924.
———. "Whaling Not What It Was." February 17, 1895.
NMDL 1797–1804 Whaling Vessels. https://nmdl.org/projects/aowv.
NMDL vessels. https://nmdl.org/projects/aowv/aowv.
NMDL Voyage ID 100198 and 100201. https://nmdl.org.
Norton, Charles C. *Journal of a Voyage to the Pacific Ocean in the Ship CANTON*. September 19, 1836. No. 355, Kendall Whaling Museum, Sharon, MA.
*Official Records of the Union and Confederate Navies in the War of the Rebellion*. Washington, D.C.: Government Printing Office, 1894–1922.
"Operation Look-See, Greenland 1957." USAF Helicopter Pilot Association. http://www.usafhpa.org.
"Overview of American Whaling: Arctic Whaling." New Bedford Whaling Museum. www.whalingmuseum.org.
Owen, Leander. "Archive of Captain Leander Owen, Legendary Whaling Master, First Steam Whaler in the Arctic." http://www.tenpound.com/171/51.html.
———. Papers of Leander Owen Stefansson Mss-247. Rauner Special Collections Library, Dartmouth College.
———. "Voyages of Owen, Leander C." National Maritime Digital Library. http://nmdl.org.
"Paloverde." Naval History and Heritage Command. https://www.history.navy.mil/research/histories/ship-histories/danfs/p/paloverde.html.
Parry, Albert. "Yankee Whalers in Siberia" *Russian Review* 5, no. 2 (Spring, 1946): 36–49.
Peck, Taylor. *Round Shots to Rockets: A History of the Washington Navy Yard and Gun Factory*. Annapolis, MD: USNI, 1949.
Philbrick, Nathaniel, "How Nantucket Came to Be the Whaling Capital of the World." *Smithsonian Magazine*, December 2015.
Pickerill, John. "American Whalers Recorded Voyages in Australian Rock Art, Study Reveals." *National Geographic*, March 1, 2019. https://www.

nationalgeographic.com/culture/2019/03/american-whalers-record-voyages-australian-rock-art.

Pinette, Megan, curator of Belfast Historical Society and Museum. Personal communication. March 2019.

"Portsmouth Fails at Whaling." http://www.seacoastnh.com/portsmouth-fails-at-whaling/?start=2.

Priolo, Gary P. "USS ATA-215." NavSource Naval History. http://www.navsource.org/archives/09/38/38215.htm.

Proctor, George. *The Fishermen's Own Book, Comprising the List of Men and Vessels Lost from the Port of Gloucester, Mass., From 1874 to April 1, 1882*. Washington, D.C.: Proctor Brothers, 1882.

"Record of American and Foreign Shipping." Mystic Seaport Museum. https://onlinebooks.library.upenn.edu/webbin/serial?id=absrecord.

"Report on the Sinking of Schooner Pilot's Bride." Federal Archives and Record Center. In *Collection of Wreck Reports, District of New London.* 1881–87.

*Republican Journal.* February 2, 1888.

Reynolds, Erminie S., and Kenneth R. Martin. *A Singleness of Purpose: The Skolfields and Their Ships*. Bath: Maine Maritime Museum, 1987.

Rice, George Wharton. *The Shipping Days of Old Boothbay from the Revolution to the World War, with Mention of Adjacent Towns*. Boothbay Harbor, ME: Southworth-Anthoensen Press, 1938.

Richards and H. Winslow. *Turnbull Library Record* 4, no. 1 (May 1971): 31–43.

Robinson, J.W. "Captain J.W. Robinson's Narrative of Sealing Voyage to Heard Island 1858–1860." Edited by W.E.L.H. Crowther. *Polar Record* 15, no. 96 (1970): 301–16.

Rogers, Erasmus Darwin. "The Journal of Erasmus Darwin Rogers, the First Man on Heard Island." Edited by R.

Ronne, Edith (Jackie). *Antarctica's First Lady: Memoirs of the First American Woman to Set Foot on the Antarctic Continent and Winter-Over*. Beaumont, TX: Beaumont Steam Museum, 2004.

———. *Transcript of Oral Interview—July 20, 2000.* Columbus, OH: Polar Archival Program. Byrd Polar Research Center. Ohio State University.

Ronne, Finn. *Antarctic Conquest: The Story of the Ronne Expedition 1946-1948.* New York: GP Putnam's Sons, 1949.

———. *Reports Covering Tests Conducted on the Ronne Antarctic Research Expedition of the American Antarctic Association, Inc. for the U.S. Air Force under Contract No. W33-038 ac-16047.* July 15, 1948.

Ronne, Finn, and Howard Liss. *The Ronne Expedition to Antarctica*. New York: Julian Messner, 1971.

Rowe, William Hutchinson. *The Maritime History of Maine: Three Centuries of Shipbuilding and Seafaring*. Gardiner, ME: Harpswell Press, 1989.

*San Francisco Call*. "Disaster, Desertion and Death." October 27, 1897.

"San Francisco Shanghaiers: 1886–1890." Mystic Seaport Museum. http://library.mysticseaport.org.

Scammon, Charles. *The Marine Mammals of the North-western Coast of North America: Together with an Account of the American Whale-fishery*. Dover, UK: 1874.

Schultz, C.R. *Inventory of the Logbooks and Journals in the G.W. Blunt White Library*. Mystic, CT: Marine Historical Association, 1965.

Semmes, Raphael. *Service Afloat; Or, the Remarkable Career of the Confederate Cruisers Sumter and Alabama*. New York: P.J. Kennedy, 1903.

Sherman, S.C. *Whaling Logbooks and Journals 1613–1927*. New York: Garland Publishing, 1986.

"Shoulder Guns." http://whalecraft.net/Shoulder_Guns.html.

Singer, Natalie. "Whaling Ship May Be Bridge's Mystery Vessel." *Seattle Times*, November 4, 2003.

Snow, Bertram G. *The Main Beam: An Informal History of Documented Vessels Built at Rockland, Maine from 1795 to 2005*. Rockland, ME: Rockland Historical Society, 2005.

"Snow Shipyards, Rockland ME." http://shipbuildinghistory.com/shipyards/small/snow.htm.

Spence, E. Lee. *Treasures of the Confederate Coast: The "Real Rhett Butler" & Other Revelations*. Charleston, SC: Narwhal Press, 1995.

———. *Shipwrecks of South Carolina and Georgia*. Sullivan's Island, SC: Sea Research Society, 1984.

"Sperm Oil Lamp, Circa 1790." Maine Memory Network. https://www.mainememory.net/artifact/16477.

Stanley, Ralph W., comp. *List of Vessels Built: Mount Desert, Cranberry, Tinker's, Thompson's and Long Island (Frenchboro)*. Searsport, ME: Stephen Phillips Memorial Library, Penobscot Marine Museum, 2003.

Starbuck, Alexander. *History of the American Whale Fishery*. Secaucus, NJ: Castle Books, 1989.

Stearns, John R. *The Story of the New England Whalers*. New York: Macmillan Co., 1908.

Stein, Douglas L. "American Maritime Documents 1776–1860." Mystic Seaport Museum Collections and Research. https://research.mysticseaport.org.

Stewardson, Jack. "Arctic Whalemen Pushed Limits of Ships, Men and Nature." *South Coast Today*. http://archive.southcoasttoday.com.

———. "Loss of the Navarch 10/27/97." *South Coast Today*. http://archive.southcoasttoday.com.

———. "Whaling Shipwreck Site May Become Historic Landmark." *South Coast Today*. August 15, 1998. http://archive.southcoasttoday.com.

Stone, Thomas. "Whalers and Missionaries at Herschel Island." *Ethnohistory* 28, no. 2 (Spring 1981): 121–24.

"Stone Fleet." World Lingo. http://www.worldlingo.com.

Summersell, Charles G. ed. *The Journal of George Townley Fullam: Boarding Officer of the Confederate Sea Raider Alabama*. Tuscaloosa: University of Alabama Press, 1973.

"*Tahmaroo* Journal 1852–1856: Log 635," Mystic Seaport Museum. http://library.mysticseaport.org.

Taylor, Nathaniel W. *Life on a Whaler or Antarctic Adventure in the Isle of Desolation*. New London, CT: New London County Historical Society, 1929.

Thaler, Andrew David. "The Menhaden of History." Southern Fried Science. May 12, 2010. http://www.southernfriedscience.com/the-menhaden-of-history.

Thompkins, E. Berkeley. "Black Ahab: William T. Shorey, Whaling Master." *California Historical Quarterly* 51, no. 1 (Spring 1972): 75–84.

Thwing, Annie Haven. *The Crooked and Narrow Streets of the Town of Boston 1630–1822*. Boston: Marshall Jones Company, 1920.

*Times*. "Sealing Ship Sinks Off Newfoundland." April 17, 1963.

"USS *American*." https://en.wikipedia.org/wiki/USS_American_(1861).

"Voyages of Amos D. Chase." National Maritime Digital Library. http://nmdl.org.

"Voyages of Leander C. Owen." National Maritime Digital Library. http://nmdl.org.

Walkerman, Sally J. "A Captain's Nightmare: The Arctic Whaling Disaster of 1871." Diss., Brown University, 2005.

*The War of the Rebellion: A Compilation of the Official Records of the Union and Confederate Armies*. Washington, D.C.: Government Printing Office, 1880–1901.

Watson, A.C., ed. "A voyage on the Sealer EMELINE." *Zoologia* 9, no. 14 (1931): 475–549.

Webb, Robert Lloyd. "Menhaden Whalemen: Nineteenth-Century Origins of American Steam Whaling." *American Neptune* 60, no. 3 (2000): 277–87.

"The Whaling Disaster of 1871." Scrimshaw by Robert Weiss. https://www.pinterest.fr/pin/524669425310182239.

"Whaling History." New Bedford Whaling Museum and Mystic Seaport Museum. https://whalinghistory.org/av/crew.

"What About Whaling?" Penobscot Marine Museum. http://www.penobscotmarinemuseum.org/pbho-1/fisheries/what-about-whaling.

Williams, Harold, ed. *One Whaling Family*. Boston: Houghton Mifflin Co., 1964.

Wise, Stephen R. *Lifeline of the Confederacy: Blockade Running in the Civil War*. Columbia: University of South Carolina Press, 1988.

## *Newspapers*

*Alexandria Gazette and Virginia Advertiser*

*Bay of Plenty Times*

*Boston Patriot and Daily Mercantile Advertiser*

*Eastern Argus*

*Honolulu Friend*

*Lincoln County News*

*National Daily Intelligencer*

*New York Times*

*Times* (London)

*Tri-Weekly Argus*

## *Personal Communication with Author*

Dodson, Robert H.T. Antarctican Society. 2018.

Good, Cipperly. Penobscot Marine Museum. September 2, 2010.

Harrison, Michael R. Nantucket Historical Association. January 31, 2019.

# ABOUT THE AUTHOR

Charles H. Lagerbom received his BA in history from Kansas State University and MA in history and archaeology from the University of Maine, with work on a Revolutionary War Truckhouse excavated on Penobscot River. An avid scuba diver, he has organized underwater surveys of ship remains in Maine lakes as well as the 1779 Penobscot Expedition. Charles worked in Antarctica with glacial geology research teams from University of Maine Quaternary Institute, now Climate Change Institute. A published author of polar, colonial Maine and maritime history, Charles has frequently written, lectured and made presentations on cruise ships, sailing vessels and ashore about history, life, politics and science of Antarctica, Cape Horn and South Atlantic, as well as colonial Maine and New England maritime history and archaeology. He is author of *The Fifth Man: The Life of H.R. Bowers*, published by Caedmon of Whitby (1999). Charles is the past membership chair of American Polar Society and immediate past president of the Antarctican Society, where he serves as current archivist/historian. He makes his home on the coast of Maine.